Universitext

Springer
*New York
Berlin
Heidelberg
Barcelona
Hong Kong
London
Milan
Paris
Singapore
Tokyo*

Universitext

Editors (North America): S. Axler, F.W. Gehring, and K.A. Ribet

(continued after index)

Joseph Rotman

Galois Theory

Second Edition

Springer

Joseph Rotman
Department of Mathematics
University of Illinois at Urbana-Champaign
Urbana, IL 61801
USA
rotman@math.uiuc.edu

Mathematics Subject Classification (2000): 12-01, 12F10

With 9 Figures.

Library of Congress Cataloging-in-Publication Data
Rotman, Joseph J., 1934-
    Galois theory / Joseph Rotman. — 2nd ed.
        p.   cm. — (Universitext)
        Includes bibliographical references (p.   –   ) and index.
        ISBN 0-387-98541-7 (softcover : alk. paper)
        1. Galois theory. I. Title.
    QA214.R685   1998
    512'.3—dc21   98-3967

Printed on acid-free paper.

Production managed by Jenny Wolkowicki; manufacturing supervised by Jacqui Ashri.
Camera-ready copy prepared from the author's AMS-TeX files.
Printed and bound by Sheridan Books, Inc., Ann Arbor, MI.
Printed in the United States of America.

9 8 7 6 5 4 3 2 (Corrected second printing)

ISBN 0-387-98541-7   SPIN 10835457

Springer-Verlag   New York Berlin Heidelberg
A member of BertelsmannSpringer Science+Business Media GmbH

To my teacher
Irving Kaplansky

Preface to the Second Edition

There are too many errors in the first edition, and so a "corrected nth print-ing" would have been appropriate. However, given the opportunity to make changes, I felt that a second edition would give me the flexibility to change any portion of the text that I felt I could improve. The first edition aimed to give a geodesic path to the Fundamental Theorem of Galois Theory, and I still think its brevity is valuable. Alas, the book is now a bit longer, but I feel that the changes are worthwhile. I began by rewriting almost all the text, trying to make proofs clearer, and often giving more details than before. Since many students find the road to the Fundamental Theorem an intricate one, the book now begins with a short section on symmetry groups of polygons in the plane; an analogy of polygons and their symme-try groups with polynomials and their Galois groups can serve as a guide by helping readers organize the various definitions and constructions. The exposition has been reorganized so that the discussion of solvability by radicals now appears later; this makes the proof of the Abel-Ruffini theo-rem easier to digest. I have also included several theorems not in the first edition. For example, the *Casus Irreducibilis* is now proved, in keeping with a historical interest lurking in these pages.

I am indebted to Gareth Jones at the University of Southampton who, after having taught a course with the first edition as text, sent me a de-tailed list of errata along with perspicacious comments and suggestions. I also thank Evan Houston, Adam Lewenberg, and Jack Shamash who made valuable comments as well. This new edition owes much to the generosity of these readers, and I am grateful to them.

Joseph Rotman
Urbana, Illinois, 1998

I thank everyone, especially Abe Seika and Bao Luong, who apprised me of errors in the first printing. I have corrected all mistakes that have been found.

Joseph Rotman
Urbana, Illinois, 2001

Preface to the First Edition

This little book is designed to teach the basic results of Galois theory—fundamental theorem; insolvability of the quintic; characterization of polynomials solvable by radicals; applications; Galois groups of polynomials of low degree—efficiently and lucidly. It is assumed that the reader has had introductory courses in linear algebra (the idea of the dimension of a vector space over an arbitrary field of scalars should be familiar) and "abstract algebra" (that is, a first course which mentions rings, groups, and homomorphisms). In spite of this, a discussion of commutative rings, starting from the definition, begins the text. This account is written in the spirit of a review of things past, and so, even though it is complete, it may be too rapid for one who has not seen any of it before. The high number of exercises accompanying this material permits a quicker exposition of it. When I teach this course, I usually begin with a leisurely account of group theory, also from the definition, which includes some theorems and examples that are not needed for this text. Here I have decided to relegate needed results of group theory to appendices: a glossary of terms; proofs of theorems. I have chosen this organization of the text to emphasize the fact that polynomials and fields are the natural setting, and that groups are called in to help.

A thorough discussion of field theory would have delayed the journey to Galois's Great Theorem. Therefore, some important topics receive only a passing nod (separability, cyclotomic polynomials, norms, infinite extensions, symmetric functions) and some are snubbed altogether (algebraic closure, transcendence degree, resultants, traces, normal bases, Kummer theory). My belief is that these subjects should be pursued only after the reader has digested the basics.

My favorite expositions of Galois theory are those of E. Artin, Kaplansky, and van der Waerden, and I owe much to them. For the appendix on "old-fashioned Galois theory," I relied on recent accounts, especially [Edwards], [Gaal], [Tignol], and [van der Waerden, 1985], and older books, especially [Dehn] (and [Burnside and Panton], [Dickson], and [Netto]). I thank my colleagues at the University of Illinois, Urbana, who, over the years, have clarified obscurities; I also thank Peter Braunfeld for suggestions that improved Appendix C and Peter M. Neumann for his learned comments on Appendix D.

I hope that this monograph will make both the learning and the teaching of Galois theory enjoyable, and that others will be as taken by its beauty as I am.

Joseph Rotman
Urbana, Illinois, 1990

To the Reader

Regard the exercises as part of the text; read their statements and do attempt to solve them all. A result labeled Theorem 1 is the first theorem in the text; Theorem G1 is the first theorem in the appendix on group theory; Theorem R1 is the first theorem in the appendix on ruler-compass constructions; Theorem H1 is the first theorem in the appendix on history.

Contents

Galois Theory

Galois theory is the interplay between polynomials, fields, and groups. The quadratic formula giving the roots of a quadratic polynomial was essentially known by the Babylonians. By the middle of the sixteenth century, the cubic and quartic formulas were known. Almost three hundred years later, Abel (1824) proved, using ideas of Lagrange and Cauchy, that there is no analogous formula (involving only algebraic operations on the coefficients of the polynomial) giving the roots of a quintic polynomial (actually Ruffini (1799) outlined a proof of the same result, but his proof had gaps and it was not accepted by his contemporaries). In 1829, Abel gave a sufficient condition that a polynomial (of any degree) have such a formula for its roots (this theorem is the reason that, nowadays, commutative groups are called abelian). Shortly thereafter, Galois (1831) invented groups, associated a group to each polynomial, and used properties of this group to give, for any polynomial, a necessary and sufficient condition that there be a formula of the desired kind for its roots, thereby completely settling the problem. We prove these theorems here.

Symmetry

Although Galois invented groups because he needed them to describe the behavior of polynomials, we realize today that groups are the precise way to describe symmetry. The Greek roots of the word *symmetry* mean, roughly, measuring at the same time. In ordinary parlance, there are at least two meanings of the word, both involving an arrangement of parts somehow balanced with respect to the whole and to each other. One of these meanings attributes an aesthetic quality to the arrangement, implying that sym-

1

metry is harmonious and well-proportioned. This usage is common in many discussions of art, and one sees it in some mathematics books as well (e.g., Weyl's *Symmetry*). Here, however, we focus on arrangements without considering, for example, whether a square is more pleasing to the eye than a rectangle.

Before giving a formal definition of symmetry, we first consider mirror images.

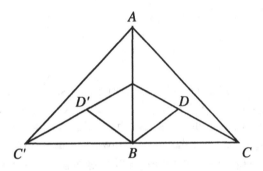

Figure 1

Let F denote the figure pictured in Figure 1. If one regards the line AB as a mirror, then the left half of F is the reflection of the right half. This figure is an example of *bilateral symmetry*: each point P on one side of AB corresponds to a point P' (its mirror image) on the other side of AB; for example, C' corresponds to C and D' corresponds to D. We can describe this symmetry in another way. Regard the plane $\mathbb{R}^2$ as a flat transparent surface in space, having F (without the letters) drawn on it. Imagine turning over this surface by flipping it around the axis AB. If one's eyes were closed before the flip and then reopened after it, one could not know, merely by looking at F in its new position, whether the flip had occurred. Indeed, if F lies in the plane so that AB lies on the y-axis and CC' lies on the x-axis, then the linear transformation $r : \mathbb{R}^2 \to \mathbb{R}^2$, defined by $(x, y) \mapsto (-x, y)$ and called a *reflection*, carries the figure into itself; that is,

$$r(F) = F.$$

On the other hand, if T is some scalene triangle in the plane (say, with its center at the origin), then it is easy to see that there are points P in T whose mirror images $P' = r(P)$ do not lie in T; that is, $r(T) \neq T$.

Another type of symmetry is *rotational symmetry*. Picture an equilateral triangle Δ in the plane with its center at the origin. A (counterclockwise) rotation ρ by 120° carries Δ into itself; if one's eyes were closed before ρ takes place and then reopened, one could not detect that a motion had occurred.

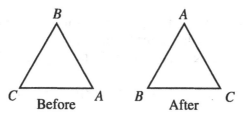

Figure 2

If we identify the plane with the complex numbers $\mathbb{C}$, then the rotation $\rho : \mathbb{C} \to \mathbb{C}$ can be described by $\rho : re^{i\theta} \mapsto re^{i(\theta + 2\pi/3)}$, and

$$\rho(\Delta) = \Delta.$$

Definition. A linear transformation $\sigma : \mathbb{R}^2 \to \mathbb{R}^2$ is called **orthogonal** if it is distance preserving; that is, if $|U - V|$ denotes the distance between points U and V, then

$$|\sigma(U) - \sigma(V)| = |U - V|.$$

There are distance preserving functions that are not linear transformations; for example, a **translation** is defined by $(x, y) \mapsto (x + a, y + b)$ for fixed numbers a and b; geometrically, this translation sends any vector (x, y) into $(x, y) + (a, b)$. (It is a theorem that every distance preserving function is a composite of reflections, rotations, and translations and, if it fixes the origin, then it is a composite of reflections and rotations alone.)

It can be shown that every orthogonal transformation σ is a bijection,[1] so that its inverse function σ^{-1} exists; moreover, one can prove that σ^{-1} is also orthogonal. The set $O(2, \mathbb{R})$ of all orthogonal transformations is a group under composition, called the *real orthogonal group*.

---

[1] A function $f : X \to Y$ is an **injection** (one also says that f is *one-to-one*) if distinct points have distinct images; that is, if $x \neq x'$, then $f(x) \neq f(x')$; the contrapositive, $f(x) = f(x')$ implies $x = x'$, is often the more useful statement. A function f is a **surjection** (one also says f is *onto*) if, for each $y \in Y$, there exists $x \in X$ with $f(x) = y$. A function f is a **bijection** (one also says f is a *one-to-one correspondence*) if it is both an injection and a surjection. Finally, a function $f : X \to Y$ is a bijection if and only if it has an **inverse**; that is, there is a function $g : Y \to X$ with both composites gf and fg identity functions.

Lemma 1. *Every orthogonal transformation σ preserves angles: if A, V and B are points, then $\angle AVB = \angle A'V'B'$, where $A' = \sigma(A)$, $V' = \sigma(V)$, and $B' = \sigma(B)$.*

Proof. We begin by proving the special case when V is the origin O. First, identify a point X with the vector starting at O and ending at X. Recall the formula relating lengths and dot product: $|X|^2 = (X, X)$, so that

$$|A - B|^2 = (A - B, A - B) = |A|^2 - 2(A, B) + |B|^2.$$

There is a similar equation for A' and B'. Since, by hypothesis, $|A' - B'| = |A - B|$, $|A'| = |A|$, and $|B'| = |B|$, it follows that $(A', B') = (A, B)$. But $(A, B) = |A||B|\cos\theta$, where $\theta = \angle AOB$. Therefore, $\angle AOB = \angle A'OB'$. But $O' = \sigma(O) = O$, because σ is a linear transformation, and so $\angle A'OB' = \angle A'O'B'$, as desired.

Now consider $\angle AVB$, where V need not be the origin O. If $\tau : W \mapsto W - V$ is the translation taking V to the origin, and if $\tau' : W \mapsto W + \sigma(V)$ is the translation taking the origin to $\sigma(V) = V'$, then the composite $\tau'\sigma\tau$ takes

$$W \mapsto W - V \mapsto \sigma(W - V) = \sigma(W) - \sigma(V) \mapsto$$
$$\sigma(W) - \sigma(V) + \sigma(V) = \sigma(W).$$

Thus, $\sigma(W) = \tau'\sigma\tau(W)$ for all W, so that $\sigma = \tau'\sigma\tau$. Since the translations τ and τ' preserve all angles, not merely those with vertex at the origin, the composite preserves $\angle AVB$. ●

The following definition of a *symmetry*, a common generalization of reflections and rotations, should now seem natural.

Definition. Given a figure F in the plane,[2] its **symmetry group** $\Sigma(F)$ is the family of all orthogonal transformations $\sigma : \mathbb{R}^2 \to \mathbb{R}^2$ for which

$$\sigma(F) = F.$$

The elements of $\Sigma(F)$ are called **symmetries**.

---

[2]It is clear that these definitions can be generalized: for every $n \geq 1$, there is an n–dimensional real orthogonal group $O(n, \mathbb{R})$ consisting of all the distance preserving linear transformations of $\mathbb{R}^n$, and symmetry groups of figures in higher dimensional euclidean space are defined as for planar figures.

It is easy to prove that the symmetry group is a subgroup of the orthog-onal group, and so it is a group in its own right.

The wonderful idea of Galois was to associate to each polynomial $f(x)$ a group, nowadays called its *Galois group*, whose properties reflect the behavior of $f(x)$. Our aim in this section is to set up an analogy between the symmetry group of a polygon and the Galois group of a polynomial.

Since our major interest is the Galois group, we merely state the fact that if σ is orthogonal and if U and V are points, then the image of the line segment UV is also a line segment, namely, $U'V'$, where $U' = \sigma(U)$ and $V' = \sigma(V)$. (The basic idea of the proof is a sharp form of the triangle inequality: if W is a point on the line segment UV, then $|UW| + |WV| = |UV|$, while if $W \notin UV$, then $|UW| + |WV| > |UV|$.)

Lemma 2. *If P is a (2-dimensional) polygon, then every orthogonal trans-formation $\sigma \in \Sigma(P)$ permutes* Vert(P), *the set of vertices of P.*

Proof. Let V be a vertex of P; if M, V, and N are consecutive vertices, then $\angle MVN \neq 180°$. If $V' = \sigma(V)$, then either V' lies on the perimeter of P or it lies in the interior of P.

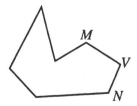

Figure 3

In the first case, Lemma 1 gives $\angle MVN = \angle M'V'N'$. But if V' is not a vertex, then $\angle M'V'N' = 180°$, a contradiction. Therefore, V' must be a vertex in this case.

In the second case, V' lies inside of P, and so there is a (2-dimensional) disk D with center V' lying wholly inside of P. Since $\sigma(P) = P$, ev-ery point in D lies in the image of σ. Now σ^{-1} is also an orthogonal transformation, and $\sigma^{-1}(V') = V$. In the disk D, every angle between $0°$ and $360°$ arises as $\angle JV'K$ for some points J and K in D. Now $\sigma^{-1}(\angle JV'K) = \angle J'VK'$ for some points J' and K' in P. But the only such angles satisfy

$$0 \leq \angle J'VK' \leq \angle MVN.$$

Therefore, there are angles that the orthogonal transformation σ^{-1} does not preserve, and this is a contradiction.

We conclude, for every vertex V, that $\sigma(V)$ is also a vertex; that is, the restriction σ_1 of σ maps Vert(P) to itself. Since σ is an injection, so is its restriction σ_1; since Vert(P) is finite, σ_1 must also be a bijection. Thus, if Vert$(P) = \{V_1, \ldots, V_n\}$, then

$$\{V_1, \ldots, V_n\} = \{\sigma(V_1), \ldots, \sigma(V_n)\} = \{\sigma_1(V_1), \ldots, \sigma_1(V_n)\},$$

and so σ_1 is a permutation of Vert(P).   •

Theorem 3. *If P is a polygon with n vertices* Vert$(P) = \{V_1, \ldots, V_n\}$, *then $\Sigma(P)$ is isomorphic to a subgroup of the symmetric group S_n.*

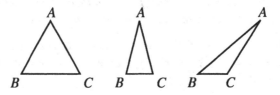

Figure 4

Proof. If $\sigma \in \Sigma(P)$, denote its restriction to Vert(P) by σ_1. By the lemma, σ_1 is a permutation of Vert(P); that is, $\sigma_1 \in S_{\text{Vert}(P)}$. It follows that the assignment $\sigma \mapsto \sigma_1$ is a well defined function $f : \Sigma(P) \to S_{\text{Vert}(P)}$.

To see that f is a homomorphism, suppose that $\sigma, \tau \in \Sigma(P)$. It is easy to see that if $V \in$ Vert(P), then $(\sigma\tau)_1$ and $\sigma_1\tau_1$ both have the same value on V, namely, $\sigma(\tau(V))$. Therefore, $(\sigma\tau)_1 = \sigma_1\tau_1$, and so f is a homomorphism:

$$f(\sigma\tau) = f(\sigma)f(\tau).$$

Finally, f is an injection, i.e., ker $f = 1$, for if $f(\sigma) = \sigma_1 = 1$, then σ fixes every vertex $V \in$ Vert(P). But regarding the vertices as vectors in $\mathbb{R}^2$, there are two such that are linearly independent (neither is a scalar multiple of the other), and so these two vectors comprise a basis of $\mathbb{R}^2$. Since σ is a linear transformation fixing a basis of $\mathbb{R}^2$, it must be the identity. Therefore, f is an isomorphism between $\Sigma(P)$ and a subgroup of $S_{\text{Vert}(P)} \cong S_n$.   •

Corollary 4. *Let Δ be a triangle with vertices A, B, and C. If Δ is equilateral, then $\Sigma(\Delta) \cong S_3$; if Δ is only isosceles, then $\Sigma(\Delta) \cong \mathbb{Z}_2$; if Δ is scalene, then $\Sigma(\Delta)$ has order 1.*

Proof. By the Theorem, $\Sigma(\Delta)$ is isomorphic to a subgroup of S_3. If Δ is equilateral, then we can exhibit 6 symmetries of it: the reflections about any of the 3 altitudes and the rotations of $0°$, $120°$ and $240°$. Since $|S_3| = 6$, it follows that $\Sigma(\Delta) \cong S_3$. If Δ is isosceles, say, $|AC| = |AB|$, then the reflection about the altitude through A is in $\Sigma(\Delta)$. This is the only non-identity symmetry, for every symmetry σ must fix A because the angle at A is different than the angles at B and C (lest Δ be equilateral). Thus, $\Sigma(\Delta) \cong \mathbb{Z}_2$. Finally, if Δ is scalene, then any symmetry fixes all the vertices, for no two angles are the same, and hence it is the identity. •

We shall see later that the Galois group of a polynomial having n distinct roots is also isomorphic to a subgroup of S_n. Moreover, there may be permutations of the roots that do not arise from the Galois group, just as there may be permutations of the vertices that do not arise from symmetries; for example, in Corollary 4 we saw that only two of the six permutations of the vertices of an isosceles triangle arise from symmetries.

Exercises

1. (i)   If F is a square, prove that $\Sigma(F) \cong D_8$, the dihedral group of order 8.

    (ii) If F is a rectangle that is not a square, prove that $\Sigma(F) \cong V$, where V denotes the 4-group ($V \cong \mathbb{Z}_2 \times \mathbb{Z}_2$).

    (iii) Give an example of quadrilaterals Q and Q' with $\Sigma(Q) \cong \mathbb{Z}_2$ and $\Sigma(Q') = 1$.

2. A polygon is *regular* if all the angles at its vertices are equal. Prove that a polygon P is regular if and only if $\Sigma(P)$ acts transitively on Vert(P).

3. Prove that if P_n is a regular polygon with n vertices, then $\Sigma(P_n) \cong D_{2n}$, where D_{2n} is the dihedral group of order $2n$.

4. Prove that if F is a circular disk, then $\Sigma(F)$ is infinite.

Rings

The algebraic system encompassing fields and polynomials is a commutative ring with 1. We assume that the reader has, at some time, heard the words *group*, *ring*, and *homomorphism*; our discussion is, therefore, not leisurely, but it is complete.

Definition. A *commutative ring with* 1 is a set R equipped with two binary operations, *addition*: $(r, r') \mapsto r + r'$ and *multiplication*: $(r, r') \mapsto rr'$, such that:

(i) R is an abelian group under addition;

(ii)   multiplication is commutative and associative;

(iii)   there is an element $1 \in R$ with $1 \neq 0$ and

$$1r = r \quad \text{for all } r \in R;$$

(iv)   the *distributive law* holds:

$$r(s + t) = rs + rt \quad \text{for all } r, s, t \in R.$$

The *additive group* of R is the abelian group obtained from it by forgetting its multiplication.

From now on, we will write *ring* instead of "commutative ring with 1."

Example 1. The most familiar rings are $\mathbb{Z}$ (the integers), $\mathbb{Q}$ (the rational numbers), $\mathbb{R}$ (the real numbers), and $\mathbb{C}$ (the complex numbers); each of them is equipped with the usual addition and multiplication.

Example 2. For a fixed positive integer n, define the ring $\mathbb{Z}_n$ of *integers modulo n* as follows. Its elements are the subsets of $\mathbb{Z}$

$$\begin{aligned} [a] &= \{m \in \mathbb{Z} : m \equiv a \bmod n\} \\ &= \{m \in \mathbb{Z} : m = a + kn \text{ for some } k \in \mathbb{Z}\}, \end{aligned}$$

where $a \in \mathbb{Z}$ ($[a]$ is called the *congruence class* of $a \bmod n$). Addition and multiplication are given by

$$[a] + [b] = [a + b] \quad \text{and} \quad [a][b] = [ab],$$

and [1] is "one." It is routine to check that addition and multiplication are well defined (that is, if $a \equiv a' \bmod n$ and $b \equiv b' \bmod n$, then $a + b \equiv a' + b' \bmod n$ and $ab \equiv a'b' \bmod n$; i.e., $[a] + [b] = [a'] + [b']$ and $[a][b] = [a'][b']$), and that $\mathbb{Z}_n$ is a ring under these operations.

Recall that $\mathbb{Z}_n$ has exactly n elements, namely

$$\mathbb{Z}_n = \{ [0], [1], \dots, [n-1] \},$$

for if $a \in \mathbb{Z}$, the division algorithm provides a quotient q and a remainder r with $a = qn + r$, where $0 \leq r < n$; it follows that $a \equiv r \bmod n$, and so $[a] = [r]$. One can also prove that the congruence classes $[r]$ for r in the indicated range are all distinct.

It is a common practice, when working within $\mathbb{Z}_n$, to eliminate the brackets from the notation. In $\mathbb{Z}_3$, for example, it is correct to write $2 + 2 = 1$.

Example 3. If R is a ring, define a *polynomial $f(x)$ with coefficients* in R (briefly, a *polynomial over R*) to be a sequence

$$f(x) = (c_0, c_1, \dots, c_n, 0, 0, \dots)$$

with $c_i \in R$ for all i and $c_i = 0$ for all $i > n$. If $g(x) = (d_0, d_1, \dots)$ is another polynomial over R, it follows that $f(x) = g(x)$ if and only if $c_i = d_i$ for all i. Denote the set of all such polynomials by $R[x]$, and define addition and multiplication on $R[x]$ as follows:

$$(c_0, c_1, \dots) + (d_0, d_1, \dots) = (c_0 + d_0, c_1 + d_1, \dots)$$

and

$$(c_0, c_1, \dots)(d_0, d_1, \dots) = (e_0, e_1, \dots),$$

where $e_0 = c_0 d_0$, $e_1 = c_0 d_1 + c_1 d_0$, and, in general, $e_k = \sum c_i d_j$, the summation being over all i, j with $i + j = k$. Define the *zero polynomial* to be $(0, 0, \dots)$, and denote it by 0; similarly, denote $(1, 0, 0, \dots)$ by 1 (there are now two meanings for these symbols). It is routine but tedious to verify that $R[x]$ is a ring, the *polynomial ring over R*.

What is the significance of the letter x in the notation $f(x)$? Let x denote the specific element of $R[x]$:

$$x = (0, 1, 0, 0, \dots).$$

It is easy to prove that $x^2 = (0, 0, 1, 0, 0, \dots)$ and, by induction, that x^i is the sequence having 0 everywhere except for 1 in the ith spot. We now recapture the usual notation:

$$
\begin{aligned}
f(x) &= (c_0, c_1, \dots, c_n, 0, 0, \dots) \\
&= (c_0, 0, 0, \dots) + (0, c_1, 0, 0, \dots) + (0, 0, c_2, 0, 0, \dots) + \cdots \\
&= c_0(1, 0, 0, \dots) + c_1(0, 1, 0, 0, \dots) + c_2(0, 0, 1, 0, \dots) + \cdots \\
&= c_0 + c_1 x + \cdots + c_n x^n \\
&= \sum c_i x^i.
\end{aligned}
$$

We have written $c_0 = c_0 1$ after identifying c_0 with $(c_0, 0, 0, \ldots)$ in $R[x]$. Notice that x is an honest element of a ring and not a variable; its role as a variable, however, will be given when we discuss polynomial functions.

We remind the reader of the usual vocabulary associated with $f(x) = c_0 + c_1 x + \cdots + c_n x^n$. If $f(x)$ is not the zero polynomial, its **leading coefficient** is c_n, where n is the largest integer with $c_n \neq 0$; one calls n the **degree** and denotes it by $\partial(f)$ [n is the highest exponent of x occurring in $f(x)$]. A **monic** polynomial is one whose leading coefficient is 1. The zero polynomial $0 = (0, 0, \ldots)$ does not have a degree, for it has no leading coefficient. The **constant term** of $f(x)$ is c_0; a **constant** (polynomial) is either the zero polynomial 0 or a polynomial of degree 0; **linear, quadratic, cubic, quartic** (or **biquadratic**), and **quintic** polynomials have degrees, respectively, 1, 2, 3, 4, and 5.

Definition. Let $f(x) = \sum c_i x^i$ be a polynomial over a ring R. A **root** of $f(x)$ in R is an element $\alpha \in R$ such that

$$c_0 + c_1 \alpha + \cdots + c_n \alpha^n = 0.$$

Remark. The polynomial $f(x) = x^2 - 2$ is a polynomial over $\mathbb{Q}$, but we usually say that $\sqrt{2}$ is a root of $f(x)$ even though $\sqrt{2}$ is irrational. We will soon modify the definition of root of a polynomial $f(x)$ over R to allow roots to lie in some ring larger than R.

Recall from linear algebra that a *linear homogeneous system* over a field with r equations in n unknowns has a nontrivial solution if $r < n$; if $r = n$, one must examine a determinant. If $f(x) = (x - \alpha_1) \ldots (x - \alpha_n) = \sum c_i x^i$, then it is easy to see, by induction on n, that

$$c_{n-1} = -\sum_i \alpha_i$$

$$c_{n-2} = \sum_{i<j} \alpha_i \alpha_j$$

$$c_{n-3} = -\sum_{i<j<k} \alpha_i \alpha_j \alpha_k$$

$$\vdots$$

$$c_0 = (-1)^n \alpha_1 \ldots \alpha_n.$$

The problem of finding the roots $\alpha_1, \ldots, \alpha_n$ of the polynomial $f(x)$ from its coefficients $c_1, \ldots, c_n$ is thus a question of solving a nonlinear system of n equations in n unknowns. We shall see that this problem is not "solvable by radicals" if $n \geq 5$.

Theorem 5. *Let R be a ring.*
  (i) *The "one" in R is unique.*
  (ii) $0 \cdot r = 0$ *for every $r \in R$;*
  (iii) *If $-r$ is the additive inverse of $r \in R$, that is, $-r + r = 0$, then*

$$-r = (-1)r;$$

  (iv) $(-1)(-r) = r$ *for every $r \in R$ [in particular, $(-1)(-1) = 1$].*

Proof. (i) Suppose that $e \in R$ satisfies $er = r$ for all $r \in R$. In particular, when $r = 1$, we have $e1 = 1$. But the defining property of 1 gives $e1 = e$, and so $e = 1$.

(ii) The distributive law gives

$$0 \cdot r = (0 + 0) \cdot r = 0 \cdot r + 0 \cdot r,$$

and subtracting $0 \cdot r$ from both sides gives $0 \cdot r = 0$.

(iii) $0 = 0 \cdot r = (-1 + 1)r = (-1)r + r$; now add $-r$ to both sides of the equation.

(iv)

$$
\begin{aligned}
0 = 0 \cdot (-r) &= (-1 + 1)(-r) \\
&= (-1)(-r) - r.
\end{aligned}
$$

Now add r to both sides   •

Suppose we do not insist, in the definition of ring, that $1 \neq 0$. If R is a "ring" in which $1 = 0$ and if $r \in R$, then

$$r = 1r = 0 \cdot r = 0;$$

hence R consists of exactly one element, namely, 0. This algebraic system is not very interesting, and so we do not consider it as a bona fide ring.

We can now see, in any ring R, why "dividing by zero" is forbidden. If $a, b \in R$, then a/b, should it exist, is an element of R such that

$$b(a/b) = a;$$

after all, dividing by b is the operation inverse to multiplying by b. In particular, if $1/0$ exists, then it is an element of R with $0 \cdot (1/0) = 1$. But $0 \cdot (1/0) = 0$, by Theorem 5(i), and this forces $1 = 0$, contrary to the inequality in the definition of ring.

Definition. A *subring* of a ring R is a subset S of R which contains 1 and which is closed under subtraction and multiplication.

For example, $\mathbb{Z}$ is a subring of $\mathbb{Q}$ which, in turn, is a subring of $\mathbb{R}$, which is a subring of $\mathbb{C}$. If R is a ring, then $R' = \{(r, 0, 0, \ldots) : r \in R\}$ is easily seen to be a subring of $R[x]$. One usually identifies R' with R; once this is done, the string of subrings above can be lengthened: $\mathbb{C}$ is a subring of $\mathbb{C}[x]$.

Exercises

5. Show that the intersection of any family of subrings of R is a subring.

6. Prove that the *binomial theorem* holds in any ring R: if $n \geq 1$, then

$$(a + b)^n = \sum \binom{n}{i} a^i b^{n-i},$$

where $\binom{n}{i}$ denotes the binomial coefficient $n!/i!(n-i)!$. (Hint: First prove that

$$\binom{n-1}{i-1} + \binom{n-1}{i} = \binom{n}{i}.)$$

7. If p is a prime, prove that p is a divisor of $\binom{p}{i}$ for $i \neq 0$ and $i \neq p$. (Note that 4 is not a divisor of $\binom{4}{2} = 6$.)

8. If R is any ring and $f(x) \in R[x]$, say, $f(x) = r_0 + r_1 x + \ldots + r_n x^n$, define its *derivative*[3] by

$$f'(x) = r_1 + 2r_2 x + \ldots + nr_n x^{n-1}.$$

Prove that

$$[f(x) + g(x)]' = f'(x) + g'(x)$$

and

$$[f(x)g(x)]' = f(x)g'(x) + f'(x)g(x).$$

---

[3]There is no notion of limit in most rings, and so we are taking the usual formula from calculus and using it to define derivative over arbitrary rings.

9. If R is a ring and S is a set, let R^S denote the set of all functions $S \to R$. Equip R^S with the operations of pointwise addition and multiplication; that is, if $f, g : S \to R$, then

$$f + g : s \mapsto f(s) + g(s),$$

and

$$fg : s \mapsto f(s)g(s).$$

Prove that R^S is a ring. (Hint. "Zero" is the constant function z with $z(s) = 0$ for all $s \in S$, and "one" is the constant function e with $e(s) = 1$ for all $s \in S$.)

Domains and Fields

Two types of ring are especially important: domains and fields.

Definition. A ring R is a **domain** (or *integral domain*) if the product of any two nonzero elements in R is itself nonzero.

Example 4. Note that $\mathbb{Z}_6$ is not a domain because $[2] \neq 0$ and $[3] \neq 0$, but $[2][3] = [6] = 0$.

Theorem 6. *A ring R is a domain if and only if it satisfies the **cancellation law**: if $ra = rb$ and $r \neq 0$, then $a = b$.*

Proof. Assume that R is a domain, that $r \neq 0$, and that $ra = rb$. Then $r(a - b) = 0$. Since R is a domain, the inequality $a - b \neq 0$ is untenable, and hence $a - b = 0$ and $a = b$.

Conversely, assume that the cancellation law holds in R. Suppose there are nonzero elements r and a in R with $ra = 0$; then $ra = 0 = r0$ implies $a = 0$, a contradiction. •

Example 4 can be generalized.

Theorem 7. $\mathbb{Z}_n$ *is a domain if and only if n is prime.*

Proof. If n is not a prime, then it is composite, and so there is a factorization $n = ab$ with $1 < a < n$ and $1 < b < n$. It follows that $[a][b] = [ab] = [n] = 0$ while $[a] \neq 0$ and $[b] \neq 0$, and so $\mathbb{Z}_n$ is not a domain.

We claim, conversely, that if p is prime, then $\mathbb{Z}_p$ is a domain. If $[a][b] = 0$ in $\mathbb{Z}_p$, then $ab \equiv 0 \bmod p$, and so p is a divisor of ab. By Euclid's lemma, which applies because p is prime, either p is a divisor of a or p is a divisor of b; that is, $[a] = 0$ or $[b] = 0$. •

There is a special name for elements in a ring that have a multiplicative inverse.

Definition. An element $u \in R$ is a *unit* if there exists $v \in R$ with $uv = 1$.

Notice that 2 is not a unit in $\mathbb{Z}$; of course, $2 \cdot \frac{1}{2} = 1$, but $\frac{1}{2} \notin \mathbb{Z}$. On the other hand, 2 is a unit in $\mathbb{Q}$.

We now define the class of rings in which one can divide by any nonzero element. Remember that dividing by s is the same as multiplying by its reciprocal s^{-1}; that is, $r \div s = rs^{-1}$, so that division by units is always possible.

Definition. A *field* is a ring R in which every nonzero $r \in R$ is a unit; that is, there is $s \in R$ with $rs = 1$.

The only units in $\mathbb{Z}$ are 1 and -1. At the other extreme, a ring R is a field if and only if every nonzero element in R is a unit.

Example 5. The rings $\mathbb{Q}$, $\mathbb{R}$, and $\mathbb{C}$ are fields.

Theorem 8. *If p is a prime, then $\mathbb{Z}_p$ is a field.*

Proof. Suppose that $[a] \in \mathbb{Z}_p$. If $[a] \neq 0$, then the integer a is not divisible by p. We claim that the gcd $(a, p) = 1$. Since p is prime, its only (positive) divisors are 1 and p, and hence 1 and p are the only candidates for the gcd; as p is not a divisor of a, however, the gcd must be 1. It follows that 1 is a linear combination of a and p: there are integers s and t with

$$1 = sa + tp.$$

Thus, $sa - 1 = -tp$, so that $sa \equiv 1 \bmod p$. In symbols,

$$[1] = [sa] = [s][a].$$

Therefore, the multiplicative inverse of $[a]$ is $[s]$, and so $\mathbb{Z}_p$ is a field.   •

Every field is a domain, for if $ra = rb$ and $r \neq 0$, then r^{-1} exists, hence $r^{-1}ra = r^{-1}rb$, and $a = b$. The converse is false; there are domains that are not fields. For example, $\mathbb{Z}$ is a domain that is not a field ($\frac{1}{3}$ is not an integer!).

Note that every subring R of a field F is a domain. After all, if r and s lie in R, then they also lie in F. If $rs = 0$ with $r \neq 0$ and $s \neq 0$, then we would contradict what we have just proved above: every field is a domain.

We are now going to prove that every domain can be viewed as a subring of a field.

Theorem 9. *For every domain R, there is a field* Frac(R) *containing R as a subring. Moreover, every element q* $\in$ Frac(R) *has a factorization*

$$q = ab^{-1}$$

with a, b $\in$ *R and b* $\neq$ 0.

Proof. We merely sketch the proof, for Frac(R) is constructed from R in exactly the same way as the field $\mathbb{Q}$ is constructed from $\mathbb{Z}$. In more detail, let X denote the set of all ordered pairs $(a, b) \in R \times R$ with $b \neq 0$, and define a "cross multiplication" relation on X:

$$(a, b) \sim (c, d) \text{ if } ad = bc.$$

This is an equivalence relation (one uses the cancellation law in proving transitivity). Denote the equivalence class of (a, b) by a/b, and denote the set $\{a/b : a, b \in R \text{ and } b \neq 0\}$ by Frac(R).

Define addition and multiplication on Frac(R) by

$$a/b + c/d = (ad + bc)/bd$$

and

$$(a/b)(c/d) = ac/bd$$

(note that $bd \neq 0$ because R is a domain). It is straightforward to check that these operations are well defined [that is, they do not depend on the choices of representative: if $a'/b' = a/b$ and $c'/d' = c/d$, then $a'/b' + c'/d' = a/b+c/d$ and $(a'/b')(c'/d') = (a/b)(c/d)$]. It is also routine to check that Frac(R) is a field; i.e., that all the field axioms do hold. In particular, one can prove that if $a, b \in R$ are both nonzero, then the inverse of a/b is b/a.

If we identify $a \in R$ with the "fraction" $a/1$ (as one identifies an integer n with the fraction $n/1$), then R can be viewed as a subring of Frac(R).

Finally, if $q \in$ Frac(R), then $q = a/b = a(1/b) = ab^{-1}$, as desired. $\bullet$

Definition. If R is a domain, then Frac(R) is called its *fraction field*.

It is easy to see that $\mathbb{Q} = $ Frac($\mathbb{Z}$). If K is a field, then Frac($K[x]$) is called the field of **rational functions** over K, and it is denoted by

$$\text{Frac}(K[x]) = K(x).$$

The elements of $K(x)$ are, of course, of the form $f(x)/g(x)$, where $f(x)$ and $g(x)$ lie in $K[x]$ and $g(x) \neq 0$ [i.e., $g(x)$ is not the zero polynomial].

Exercises

10.   (i) If R is a ring, prove that $U(R)$, the set of all units in R, is a group under multiplication. One calls $U(R)$ the *group of units* of R.

(ii) Prove that a ring R is a field if and only if $R^{\#} \doteq R - 0$ is a group under multiplication. (Of course, $U(R) = R^{\#}$ here.)

11.   Prove that if $a \in \mathbb{Z}$, then $[a]$ is a unit in $\mathbb{Z}_n$ if and only if $(a, n) = 1$. Conclude that the group of units, $U(\mathbb{Z}_n)$, has order $\varphi(n)$, where φ is *Euler's function*: $\varphi(1) = 1$ and, if $n > 1$, then $\varphi(n) = |\{k \in \mathbb{Z} : 1 \le k < n$ and $(k, n) = 1\}|$.

12.   Let $f(x), g(x) \in R[x]$. Show that the constant term of $f(x)g(x)$ is the product of the constant terms of $f(x)$ and of $g(x)$.

13.   (i) If R is a domain, then the leading coefficient of $f(x)g(x)$ is the product of the leading coefficients of $f(x)$ and of $g(x)$. Conclude that if $f(x)$ and $g(x)$ are nonzero polynomials in $R[x]$, where R is a domain, then
$$\partial(fg) = \partial(f) + \partial(g).$$

(ii) Prove that if R is a domain, then $R[x]$ is also a domain.

(iii) If $R = \mathbb{Z}_4[x]$, show that $(2x + 1)^2 = 1$. Conclude that the formula $\partial(fg) = \partial(f) + \partial(g)$ may be false in $R[x]$ when R is not a domain.

(iv) Show that there is a factorization $x = f(x)g(x)$ in $R = \mathbb{Z}_4[x]$ in which neither $f(x)$ nor $g(x)$ is a constant.

14.   Define the ring of *polynomials in two variables* over R, denoted by $R[x, y]$, as $A[y]$, where $A = R[x]$. Define polynomials in several variables over R by induction, and show that if R is a domain, then so is $R[x_1, \dots , x_n]$ (one usually denotes $\mathrm{Frac}(F[x_1, \dots , x_n])$ by $F(x_1, \dots , x_n)$ when F is a field).

15.   Let R be a domain, and let $f, g \in R$ be nonzero elements satisfying
$$f = ug \quad \text{and} \quad g = vf,$$
where $u, v \in R$. Prove that $uv = 1$ and that u and v are units.

16.   (i) Prove that if F is a field, then the units in $F[x]$ are the nonzero constants.

(ii) Show that $\mathbb{Z}_2[x]$ is an infinite ring having exactly one unit.

(iii) Give an example of a nonconstant polynomial in $\mathbb{Z}_4[x]$ that is a unit.

17. (i) Prove the **division algorithm** for polynomials: If R is a ring, if $f(x)$, $g(x) \in R[x]$, and if the leading coefficient of $g(x)$ is a unit [in particular, if $g(x)$ is monic], then there are polynomials $q(x)$ and $r(x) \in R[x]$ (**quotient** and **remainder**) with

$$f(x) = q(x)g(x) + r(x)$$

and either $r(x) = 0$ or $\partial(r) < \partial(g)$.

(ii) If R is a domain, then the quotient and remainder occurring in the division algorithm are unique. (There are rings R, e.g. $\mathbb{Z}_4$, for which the corresponding assertion is false.)

18. A **subfield** F of a ring R is a subring of R that is a field. Show that a subset X of a ring R is a subfield if and only if X contains 1 and X is closed under subtraction, multiplication, and inverses.

19. Prove that the intersection of any family of subfields is itself a subfield. (Note that this intersection is not $\{0\}$ because it contains 1.)

20. (i) Show that $\mathbb{Z}_p[x]$ is an infinite domain containing $\mathbb{Z}_p$ as a subfield.

(ii) Show that there exists an infinite field containing $\mathbb{Z}_p$ as a subfield.

21. Show that $R[x]$ is never a field.

22. Show that $\mathbb{Z}_n$ is a field if and only if n is prime.

Homomorphisms and Ideals

It is useful to study transformations from one ring to another.

Definition. If R and S are rings, then a function $\varphi : R \to S$ is a **ring homomorphism** (or **ring map**) if, for all $r, r' \in R$:

$$\begin{aligned}
\varphi(r + r') &= \varphi(r) + \varphi(r'); \\
\varphi(rr') &= \varphi(r)\varphi(r'); \\
\varphi(1) &= 1.
\end{aligned}$$

A ring homomorphism $\varphi : R \to S$ is an **isomorphism** if φ is a bijection; in this case, one says that R and S are **isomorphic** and one writes $R \cong S$.

The first two examples show that isomorphisms make our earlier "identifications" more precise.

Example 6. If R is a ring, then $R' = \{(r, 0, 0, \dots) : r \in R\}$ is a subring of $R[x]$, and the function $\varphi : R \to R'$, defined by

$$\varphi : r \mapsto (r, 0, 0, \dots),$$

is an isomorphism from R to R'.

Example 7. Let R be a domain with field of fractions $F = \text{Frac}(R)$. It is easy to see that $R'' = \{r/1 \in F : r \in R\}$ is a subring of F, and that

$$\varphi : r \mapsto r/1$$

is an isomorphism from R to R''.

Example 8. The map $\pi : \mathbb{Z} \to \mathbb{Z}_n$, defined by $\pi : a \mapsto [a]$, is a surjective ring map.

Definition. If $\varphi : R \to S$ is a ring map, then its **kernel** is

$$\ker \varphi = \{r \in R : \varphi(r) = 0\},$$

and its **image** is

$$\text{im } \varphi = \{s \in S : s = \varphi(r) \ \text{for some} \ r \in R\}.$$

It is easy to check that if $\varphi : R \to S$ is a ring homomorphism, then $\ker \varphi$ is an additive subgroup of R that is closed under multiplication (it is not a subring because it does not contain 1), and $\text{im } \varphi$ is a subring of S. In group theory, the kernel of a homomorphism is not merely a subgroup; it is a normal subgroup. Similarly, in ring theory, kernels have added structure.

Definition. An **ideal** in a ring R is a subset I containing 0 such that:

(i) $a, b \in I$ imply $a - b \in I$;

(ii) $a \in I$ and $r \in R$ imply $ra \in I$.

An ideal I in a ring R is a **proper ideal** if $I \neq R$.

Every ring R contains the ideals R itself and $\{0\}$.

Theorem 10. *If* $\varphi : R \to S$ *is a ring homomorphism, then* $\ker \varphi$ *is a proper ideal in* R. *Moreover,* φ *is an injection if and only if* $\ker \varphi = \{0\}$.

Proof. If one forgets the multiplications in the rings R and S and remembers only that they are additive abelian groups, then φ is a homomorphism of groups, and so $\ker \varphi$ is an additive subgroup of the additive group R. If $a \in I$ and $r \in R$, then

$$\varphi(ra) = \varphi(r)\varphi(a) = \varphi(r) \cdot 0 = 0.$$

Therefore, $ra \in \ker \varphi$, and so $\ker \varphi$ is an ideal in R. Since $\varphi(1) = 1 \neq 0$, we see that $\ker \varphi \neq R$, and so $\ker \varphi$ is a proper ideal.

If φ is an injection, then distinct points have distinct images. In particular, if $r \neq 0$, then $\varphi(r) \neq \varphi(0) = 0$, so that $r \notin \ker \varphi$, and $\ker \varphi = \{0\}$. Conversely, suppose that $\ker \varphi = \{0\}$. If $\varphi(r) = \varphi(r')$, then

$$0 = \varphi(r) - \varphi(r') = \varphi(r - r');$$

hence, $r - r' \in \ker \varphi = \{0\}$, and so $r = r'$; therefore, φ is an injection.   •

Exercises

23. If R is a field, prove that the map $R \to \mathrm{Frac}(R)$, given by $a \mapsto a/1$, is an isomorphism. Conversely, prove that if R is a domain and the map $a \mapsto a/1$ is an isomorphism, then R is a field.

24. If $\varphi : R \to S$ is an isomorphism between domains, prove that there is an isomorphism $\mathrm{Frac}(R) \to \mathrm{Frac}(S)$, namely, $a/b \mapsto \varphi(a)/\varphi(b)$.

25. Let R be a subring of a field F, and let K be the intersection of all the subfields of F that contain R. Prove that $K \cong \mathrm{Frac}(R)$.

26. (i) If $\varphi : R \to S$ is an isomorphism, then $\varphi^{-1} : S \to R$ is also an isomorphism.

    (ii) If $\varphi : R \to S$ and $\psi : S \to T$ are ring homomorphisms, then so is their composite $\psi\varphi : R \to T$.

27. If $a \in R$ is a unit in R and if $\varphi : R \to S$ is a ring map, then $\varphi(a)$ is a unit in S.

28.   (i)   If R is a ring, prove that $\varphi : R[x] \to R$, where $\varphi : f(x) \mapsto c_0$, the constant term of $f(x)$, is a ring map.

    (ii)   What is $\ker \varphi$?

29.   (i)   If $\sigma : R \to S$ is a ring map, prove that $\sigma^* : R[x] \to S[x]$, defined by

$$\sum r_i x^i \mapsto \sum \sigma(r_i) x^i,$$

is also a ring map.

    (ii)   If $\tau : S \to T$ is a ring map, prove that $(\tau \sigma)^* : R[x] \to T[x]$ is equal to $\tau^* \sigma^*$.

    (iii)   Prove that if σ is an isomorphism, then so is σ^*.

30.   (i)   The intersection of any family of ideals in R is an ideal in R. Conclude that if X is any subset of a ring R, there is a smallest ideal, denoted by (X), containing X. One calls (X) the *ideal generated by* X, namely, the intersection of all the ideals in R that contain X.

    (ii)   Prove that (X) is the "smallest" ideal containing X in the following sense: (X) is an ideal containing X and, if J is any ideal in R containing X, then $(X) \subset J$.

31.   (i)   If $a \in R$, prove that $\{ra : r \in R\}$ is the ideal generated by a; it is called the *principal ideal generated by* a, and it is denoted by (a).

    (ii)   If $a_1, \ldots, a_n$ are elements in a ring R, prove that the set of all linear combinations,

$$I = \{r_1 a_1 + \cdots + r_n a_n : r_i \in R, i = 1, \ldots, n\},$$

is equal to $(a_1, \ldots, a_n)$, the ideal generated by $\{a_1, \ldots, a_n\}$.

32.   Let u be a unit in a ring R.

    (i)   Prove that if an ideal I contains u, then $I = R$.

    (ii)   If $r \in R$, then $(ur) = (r)$. In particular, every nonzero principal ideal $(f(x))$ in $R = F[x]$, where F is a field, can be generated by a monic polynomial.

    (iii)   If R is a domain and $r, s \in R$, then $(r) = (s)$ if and only if $s = ur$ for some unit u in R.

33.   Prove that a ring R is a field if and only if it has only one proper ideal, namely, $\{0\}$.

34. (i) The set I of all $f(x) \in \mathbb{Z}[x]$ having even constant term is an ideal in $\mathbb{Z}[x]$; it consists of all the linear combinations of x and 2; that is, $I = (x, 2)$.

(ii) Prove that $(x, 2)$ is not a principal ideal in $\mathbb{Z}[x]$.

35. Prove that if F is a field and S is a ring, then a ring map $\varphi : F \to S$ must be an injection and im φ is a subfield of S isomorphic to F.

Quotient Rings

Let I be an ideal in R. Forgetting the multiplication for a moment, I is a subgroup of the additive group of R; moreover, R abelian implies that I is a normal subgroup, and so the quotient group R/I exists. The elements of R/I are the cosets $r + I$, where $r \in R$, and addition is given by

$$(r + I) + (r' + I) = (r + r') + I;$$

in particular, the zero element is $0 + I = I$. Recall that $r + I = r' + I$ if and only if $r - r' \in I$. Finally, the **natural map** $\pi : R \to R/I$ is the surjective (group) homomorphism defined by $r \mapsto r + I$.

Theorem 11. *Let I be a proper ideal in a ring R. Then the additive abelian group R/I can be equipped with a multiplication which makes it a ring and which makes the natural map $\pi : R \to R/I$ a surjective ring homomorphism.*

Proof. Define multiplication on R/I by

$$(r + I)(r' + I) = rr' + I.$$

To see that this is well defined, suppose that $r + I = s + I$ and that $r' + I = s' + I$; we must show that $rr' + I = ss' + I$; that is, $rr' - ss' \in I$. But

$$rr' - ss' = (rr' - rs') + (rs' - ss') = r(r' - s') + (r - s)s'.$$

Now $r' - s' \in I$ and $r - s \in I$, by hypothesis; hence $r(r' - s') \in I$ and $(r - s)s' \in I$, because I is an ideal. Finally, the sum of two elements of I is again in I, so that $rr' - ss' \in I$ and $rr' + I = ss' + I$, as desired.

It is routine to see that the abelian group R/I equipped with this multiplication is a ring; in particular, "one" is $1 + I$. Since I is a proper ideal, $1 \notin I$, and so $1 + I \neq 0 + I = 0$. The formula $(r + I)(r' + I) = rr' + I$ says that $\pi(r)\pi(r') = \pi(rr')$, where $\pi : r \mapsto r + I$ is the natural map. It follows that π is a surjective ring homomorphism. ●

Definition. If I is an ideal in a ring R, then R/I is called the *quotient ring* of R *modulo* I.

In Exercise 36, one sees that $\mathbb{Z}_n$ is equal to the quotient ring R/I, where $R = \mathbb{Z}$ and $I = (n)$, the principal ideal generated by n.

Example 9. Let $R = F[x]$, the polynomial ring over a field F; let I be the (principal) ideal generated by some particular polynomial $p(x)$ of degree n, so that $I = \{f(x)p(x) : f(x) \in F[x]\}$. If $g(x) \in F[x]$, then the division algorithm gives $q(x)$ and $r(x)$ in $F[x]$ with

$$g(x) = q(x)p(x) + r(x),$$

where $r = 0$ or $\partial(r) < n$. Note that $g(x) + I = r(x) + I$, so we may assume that every coset (except I itself) has a representative of degree $< n$. Indeed, each such coset has a unique representative $r(x)$ of degree $< n$: if there were a second such, say, $r'(x)$, then $r - r' \in I = (p)$, so that $p \mid r - r'$ and $r - r' = pf$ for some $f(x) \in F[x]$. But $r - r'$ has degree $< n = \partial(p)$, while $\partial(pf) \geq \partial(p)$, and this is a contradiction. The multiplication in R/I can be simplified: $(f(x) + I)(g(x) + I) = f(x)g(x) + I = r(x) + I$, where $r(x)$ is the remainder after dividing $f(x)g(x)$ by $p(x)$.

Example 10. Consider the special case of the preceding example in which $F = \mathbb{R}$ and $p(x) = x^2 + 1$. In $\mathbb{R}[x]/I$, where $I = (x^2 + 1)$, every element has the form $a + bx + I$, where $a, b \in \mathbb{R}$, for $x^2 + 1$ has degree 2. Moreover,

$$\begin{aligned}(a + bx + I)(c + dx + I) &= (a + bx)(c + dx) + I \\ &= ac + (bc + ad)x + bdx^2 + I.\end{aligned}$$

Now $x^2 \equiv -1 \bmod (p(x))$, so that

$$x^2 + I = -1 + I.$$

It follows that

$$(a + bx + I)(c + dx + I) = ac - bd + (bc + ad)x + I.$$

Now $\mathbb{R}[x]/I$ is actually a field, for it is easy to exhibit the multiplicative inverse of $a + bx + I$ (where $a \neq 0$ or $b \neq 0$), namely, $c + dx + I$, where $c = a/(a^2 + b^2)$ and $d = -b/(a^2 + b^2)$. We let the reader prove that the map $\varphi : \mathbb{R}[x]/I \to \mathbb{C}$, defined by $a + bx + I \mapsto a + bi$, is an isomorphism of fields. In particular, the "imaginary" number i with $i^2 = -1$ is equal to the coset $x + I$.

Theorem 12 (First Isomorphism Theorem). *If $\varphi : R \to S$ is a ring homomorphism with* $\ker \varphi = I$, *then there is an isomorphism* $R/I \to \operatorname{im} \varphi$ *given by* $r + I \mapsto \varphi(r)$.

Proof. If we forget the multiplication in R and in S, then the First Isomorphism Theorem for groups says that the function $\Phi : R/I \to \operatorname{im} \varphi$, defined by $\Phi : r + I \mapsto \varphi(r)$, is an isomorphism of additive groups. Since $\Phi(1 + I) = \varphi(1) = 1$, the proof will be complete if we prove that Φ preserves multiplication. Now if $r, r' \in R$, then

$$\Phi((r + I)(r' + I)) = \Phi(rr' + I) = \varphi(rr') = \varphi(r)\varphi(r'),$$

because φ is a ring map. But $\Phi(r + I)\Phi(r' + I) = \varphi(r)\varphi(r')$ as well, and so

$$\Phi((r + I)(r' + I)) = \Phi(r + I)\Phi(r' + I),$$

as desired.   •

As in group theory, there is a correspondence theorem (see Exercise 38). There are also second and third isomorphism theorems for rings, but they are less interesting than their group theoretic analogues.

Exercises

36. Let n be a positive integer and let $I = (n)$ be the principal ideal in $\mathbb{Z}$ generated by n. Show that the quotient ring $\mathbb{Z}/I$ is equal to $\mathbb{Z}_n$, the ring of integers modulo n. (Hint. These rings have the same elements ($[a] = a + I$) and the same addition and multiplication.)

37. Prove that if R is a ring and $I = (x)$ is the principal ideal in $R[x]$ generated by x, then $R[x]/I \cong R$.

38. Prove the *Correspondence Theorem for Rings*. If I is a proper ideal in a ring R, then there is a bijection from the family of all intermediate ideals J, where $I \subset J \subset R$, to the family of all ideals in R/I, given by

$$J \mapsto \pi(J) = J/I = \{a + I : a \in J\},$$

where $\pi : R \to R/I$ is the natural map. Moreover, if $J \subset J'$ are intermediate ideals, then $\pi(J) \subset \pi(J')$. (Compare with Theorem G.9.)

39. Let I be an ideal in a ring R, let J be an ideal in a ring S, and let $\varphi : R \to S$ be a ring isomorphism with $\varphi(I) = J$. Prove that the function $\overline{\varphi} : r + I \mapsto \varphi(r) + J$ is a (well defined) isomorphism $R/I \to S/J$.

Polynomial Rings over Fields

Theorem 13. *If F is a field, then every ideal in $F[x]$ is a principal ideal.*

Proof. Let I be an ideal in $F[x]$. If $I = \{0\}$, then $I = (0)$, the principal ideal generated by 0. If $I \neq \{0\}$, choose a polynomial $m(x)$ in I having smallest degree; we claim that $I = (m(x))$.

Clearly, $(m(x)) \subset I$. For the reverse inclusion, take $f(x)$ in I. By the division algorithm, there are polynomials $q(x)$ and $r(x)$ with

$$f(x) = q(x)m(x) + r(x),$$

where either $r(x) = 0$ or $\partial(r) < \partial(m)$. Now $r(x) = f(x) - q(x)m(x) \in I$; if $r(x) \neq 0$, then we have contradicted $m(x)$ having the smallest degree of all polynomials in I. Therefore $r(x) = 0$ and $f(x) = q(x)m(x) \in (m(x))$. •

By Exercise 32(ii), one may choose $m(x)$ to be monic (since F is a field).

Definition. A ring R is called a ***principal ideal domain*** (PID) if it is a domain in which every ideal is a principal ideal.

Of course, the reader knows that $\mathbb{Z}$ is a PID. Theorem 13 shows that $F[x]$ is a PID when F is a field; on the other hand, $\mathbb{Z}[x]$ is not a PID (in Exercise 34, it is shown that the ideal I in $\mathbb{Z}[x]$ consisting of all the polynomials having even constant term is not a principal ideal).

Definition. Let R be a ring; if $r, s \in R$, then r ***divides*** s (or s is a ***multiple*** of r) if there exists $r' \in R$ with $rr' = s$; one writes $r \mid s$ in this case.

Note that $r \mid s$ if and only if $s \in (r)$, the principal ideal generated by r. It is easy to see that $r \mid 0$ for every $r \in R$, but that $0 \mid r$ if and only if $r = 0$; also, $r \mid r$ for every $r \in R$, and r is a unit if and only if $r \mid 1$.

Definition. Let R be a domain, and let $f(x), g(x) \in R[x]$. The ***greatest common divisor*** (gcd) of $f(x)$ and $g(x)$ is a polynomial $d(x) \in R[x]$ such that:

(i) $d(x)$ is a common divisor of $f(x)$ and $g(x)$; that is, $d \mid f$ and $d \mid g$;

(ii) if $c(x)$ is any common divisor of $f(x)$ and $g(x)$, then $c(x) \mid d(x)$;

(iii) $d(x)$ is monic.

One often denotes $d(x)$ by (f, g). If $(f, g) = 1$, then $f(x)$ and $g(x)$ are called *relatively prime*.

Note that the gcd d of f and g, if it exists, is unique. If d' is another gcd, then regard it only as a common divisor and use (ii) to obtain $d' \mid d$; similarly, $d \mid d'$ if one regards d merely as a common divisor. By Exercise 15, $d' = ud$ for some unit $u \in F[x]$; that is, $d' = ud$ for some nonzero constant u (Exercise 32). Since d and d' are both monic, however, $u = 1$ and $d' = d$.

If a linear combination of polynomials f and g is 1, say, there are polynomials $a(x)$ and $b(x)$ with $1 = a(x)f(x) + b(x)g(x)$, then f and g must be relatively prime. After all, any common divisor $c(x)$ of f and g must also divide 1, and hence c is a unit. The next result shows that the gcd in $F[x]$, when F is a field, is always a linear combination.

Theorem 14. *Let F be a field and let $f(x), g(x) \in F[x]$. Then the gcd $(f(x), g(x)) = d(x)$ exists, and it is a linear combination of $f(x)$ and $g(x)$; that is, there are polynomials $a(x)$ and $b(x)$ with*

$$d(x) = a(x)f(x) + b(x)g(x).$$

Proof. By Exercise 31,

$$I = \{a(x)f(x) + b(x)g(x) : a(x), b(x) \in F[x]\}$$

is an ideal in $F[x]$ containing both $f(x)$ and $g(x)$. Since F is a field, $F[x]$ is a PID and I is a principal ideal. By Exercise 32, we may choose a monic polynomial $d(x)$ with $I = (d(x))$; as is every element of I, the generator d is a linear combination of f and g. Now d is a common divisor of f and g because $f, g \in I = (d)$. Finally, if c is a common divisor, then $c \mid f$ and $c \mid g$; that is, $f = cc'$ and $g = cc''$. Hence, $d = af + bg = acc' + bcc'' = c(ac' + bc'')$, and so $c \mid d$. Therefore, $d(x)$ is the gcd.   •

Example 11. Let $R = F[x]/I$, where F is a field and I is the principal ideal generated by some polynomial $p(x)$. If $f(x)$ and $p(x)$ are relatively prime, then there are polynomials $s(x), t(x) \in F[x]$ with

$$s(x)f(x) + t(x)p(x) = 1;$$

in R this equation becomes

$$s(x)f(x) + I = 1 + I.$$

Thus $f(x) + I$ is a unit in R with inverse $s(x) + I$.

The converse is also true. If $f(x) + I$ is a unit in R, then there is $g(x) \in F[x]$ with $1 + I = (f(x) + I)(g(x) + I) = f(x)g(x) + I$; that is,

$$f(x)g(x) - 1 = h(x)p(x)$$

for some $h(x) \in F[x]$. Therefore, $f(x)$ and $g(x)$ are relatively prime.

Corollary 15 (Euclid's Lemma). *Let F be a field, and let $p(x) \in F[x]$ not be a product of polynomials of smaller degree; if $p(x)$ divides a product $q_1(x) \cdots q_n(x)$, then $p(x)$ divides $q_j(x)$ for some j.*

Proof. By induction on $n \geq 2$, it suffices to prove that if $(f(x), g(x)) = 1$ and $f(x)$ divides $g(x)h(x)$, then $f(x)$ divides $h(x)$. There are polynomials $a(x)$ and $b(x)$ with $1 = af + bg$. Hence $h = afh + bgh$. By hypothesis, $gh = fk$ for some polynomial k, so that $h = afh + bfk = f(ah + bk)$ and f divides h. •

The proof of Theorem 14 yields the following fact; it also explains why the gcd of f and g is denoted by (f, g).

Corollary 16. *Let F be a field, let $f(x), g(x) \in F[x]$, and let the ideal generated by $f(x)$ and $g(x)$ be $I = (f(x), g(x))$. Then $I = (d(x))$, where $d(x)$ is the gcd of $f(x)$ and $g(x)$.*

The proof of Euclid's lemma is just an adaptation of the usual proof of Euclid's lemma in $\mathbb{Z}$; the same is true for the euclidean algorithm to be proved next. If one is given explicit polynomials $f(x)$ and $g(x)$, how can one compute their gcd? How can one express the gcd as a linear combination?

Theorem 17 (Euclidean Algorithm). *There are algorithms to compute the gcd and to express it as a linear combination.*

Proof. The idea is just to iterate the division algorithm. Consider the list of equations:

$$
\begin{array}{llll}
f & = & q_1 g + r_1 & \qquad \partial(r_1) \;<\; \partial(g) \\
g & = & q_2 r_1 + r_2 & \qquad \partial(r_2) \;<\; \partial(r_1) \\
r_1 & = & q_3 r_2 + r_3 & \qquad \partial(r_3) \;<\; \partial(r_2) \\
& \vdots & & \qquad\qquad \vdots \\
r_{n-2} & = & q_n r_{n-1} + r_n & \qquad \partial(r_n) \;<\; \partial(r_{n-1}) \\
r_{n-1} & = & q_{n+1} r_n + r_{n+1} & \qquad \partial(r_{n+1}) \;<\; \partial(r_n) \\
r_n & = & q_{n+2} r_{n+1}
\end{array}
$$

(note that all q_i and r_i are explicitly known from the division algorithm). We claim that $d = r_{n+1}$ is the gcd (after it is made monic). First of all, note that the iteration must stop because the degrees of the remainders strictly decrease [indeed, the number of steps needed is less than $\partial(g)$]. Second, d is a common divisor, for $d = r_{n+1}$ divides r_n and so the $(n+1)$st equation $r_{n-1} = q_{n+1}r_n + r_{n+1}$ shows that $d \mid r_{n-1}$. Working up the list in this way ultimately gives: $d \mid g$ and $d \mid f$. Third, if c is a common divisor, start at the top of the list and work down: $c \mid f$ and $c \mid g$ imply $c \mid r_1$; then $c \mid g$ and $c \mid r_1$ imply $c \mid r_2$; and so forth. Therefore, d is the gcd.

Finally, one finds a and b by working from the bottom of the list upward. Thus $d = r_{n+1} = r_{n-1} - q_{n+1}r_n$ is a linear combination of r_{n-1} and r_n. Combining this with $r_n = r_{n-2} - q_n r_{n-1}$ gives

$$
\begin{aligned}
d &= r_{n-1} - q_{n+1}(r_{n-2} - q_n r_{n-1}) \\
&= (1 + q_n q_{n+1})r_{n-1} - q_{n+1}r_{n-2},
\end{aligned}
$$

a linear combination of r_{n-2} and r_{n-1}. This process ends with d as a linear combination of f and g. •

In practice, the euclidean algorithm is quite tedious to implement (see Exercise 45), but it is useful if one wants to compute the multiplicative inverse of $f(x)$ in $F[x]/(p(x))$ when f and p are relatively prime (see Example 11). The next corollary is another valuable consequence of the euclidean algorithm.

Corollary 18. *Let* $k \subset K$ *be fields, and let* $f(x), g(x) \in k[x] \subset K[x]$. *Then the* gcd *of* f *and* g *computed in* $K[x]$ *is the same as the* gcd *of* f *and* g *computed in* $k[x]$.

Proof. The division algorithm in $K[x]$ gives

$$ f(x) = Q(x)g(x) + R(x), $$

where $Q(x), R(x) \in K[x]$ and $\partial(R) < \partial(g)$; since also $f(x), g(x) \in k[x]$, the division algorithm in $k[x]$ gives

$$ f(x) = q(x)g(x) + r(x), $$

where $q(x), r(x) \in k[x]$ and $\partial(r) < \partial(g)$. But the equation $f(x) = q(x)g(x) + r(x)$ also holds in $K[x]$, because $k[x] \subset K[x]$, so that the uniqueness of the quotient and remainder in the division algorithm in $K[x]$ gives $Q(x) = q(x)$ and $R(x) = r(x)$. Hence, the list of equations occurring in the euclidean algorithm in $K[x]$ is identical with the list occurring in the euclidean algorithm in the smaller ring $k[x]$. Therefore, the same gcd is obtained in both polynomial rings. •

Definition. Let F be a field, and let $f(x), g(x) \in F[x]$. The *least common multiple* (lcm) of $f(x)$ and $g(x)$ is a polynomial $m(x) \in F[x]$ such that:

(i) $m(x)$ is a common multiple of $f(x)$ and $g(x)$; that is, $f \mid m$ and $g \mid m$;

(ii) if $c(x)$ is any common multiple of $f(x)$ and $g(x)$, then $m(x) \mid c(x)$;

(iii) $m(x)$ is monic.

The next result should be compared with Corollary 16.

Theorem 19. *If F is a field and $f(x), g(x) \in F[x]$, then their lcm is the monic generator of $(f) \cap (g)$.*

Proof. Since $F[x]$ is a PID, the ideal $(f) \cap (g) = (m)$ for some monic polynomial $m(x) \in F[x]$. Now $m \in (f)$ implies $m(x) = f(x)r(x)$ for some $r(x) \in F[x]$, and $m \in (g)$ implies $m(x) = g(x)s(x)$ for some $s(x) \in F[x]$; thus, $m(x)$ is a common multiple of f and g. Finally, if $h(x)$ is another common multiple of f and g, then $h(x) = f(x)r'(x) = g(x)s'(x)$; that is, $h \in (f) \cap (g) = (m)$ and $m \mid h$. •

There is an elementary relation between factoring and roots.

Theorem 20. *Let $f(x) \in F[x]$ and let $a \in F$. Then there is $q(x) \in F[x]$ with*
$$f(x) = q(x)(x - a) + f(a).$$

Proof. By the division algorithm, there is an equation in $F[x]$:
$$f(x) = (x - a)q(x) + r(x),$$
where either $r(x) = 0$ or $\partial(r) < 1 = \partial(x - a)$; that is, $r(x)$ is a constant. Evaluating at a gives the equation $f(a) = q(a)(a - a) + r = r$. •

Corollary 21. *Let $f(x) \in F[x]$. Then $a \in F$ is a root of $f(x)$ if and only if $x - a$ divides $f(x)$.*

Proof. If a is a root of $f(x)$, then $f(a) = 0$, and the theorem gives $f(x) = q(x)(x - a)$. Conversely, if $f(x) = q(x)(x - a)$, then evaluating at a gives $f(a) = 0$ and a is a root of $f(x)$. •

Theorem 22. *If F is a field and $f(x) \in F[x]$ has degree $n \geq 0$, then F contains at most n roots of $f(x)$.*

Proof. Suppose that F contains $n + 1$ distinct roots of $f(x)$, say, $a_1, \ldots ,$ a_{n+1}. By the corollary, $f(x) = (x - a_1)g_1(x)$ for some $g_1(x) \in F[x]$. Now $x - a_2$ divides $(x - a_1)g_1(x)$; since $a_2 \neq a_1$, the polynomials $x - a_1$ and $x - a_2$ are relatively prime, and so Euclid's lemma gives $x - a_2$ dividing $g_1(x)$; hence

$$f(x) = (x - a_1)(x - a_2)g_2(x).$$

Using Exercise 42, one proves by induction on i that

$$f(x) = (x - a_1)(x - a_2) \cdots (x - a_i)g_i(x),$$

and so

$$f(x) = (x - a_1)(x - a_2) \ldots (x - a_{n+1})g_{n+1}(x).$$

This cannot be, for the left side $f(x)$ has degree n while the right side has degree greater than n.   •

The last theorem is false for arbitrary rings R; for example, $x^2 - 1$ has four roots in $\mathbb{Z}_8$, namely, [1], [3], [5], and [7].

Example 12. If $a \in R$, define $e_a : R[x] \to R$ by

$$f(x) = \sum r_i x^i \mapsto \sum r_i a^i.$$

The element $e_a(f) = \sum r_i a^i \in R$ is denoted by $f(a)$. We let the reader check that e_a is a ring map; it is called **evaluation at a**. Thus, each polynomial $f(x) \in R[x]$ determines a **polynomial function** $f : R \to R$, namely, $f : a \mapsto f(a) = e_a(f)$, and so we may now regard x as a variable ranging over R.

Example 13. It is easy to check that the map $\varphi : R[x] \to R^R$ (see Exercise 9), which assigns to each polynomial $f(x)$ its polynomial function f, is a homomorphism. It follows that $P(R)$, defined as im φ, is a subring of R^R; we call $P(R)$ the **ring of polynomial functions** over R. But here is a surprise: φ need not be an injection; ker φ can be nonzero. For example, we know that if $p \neq 1$, then $x^p \neq x$, because their coefficients do not match. On the other hand, if p is a prime, then x^p and x determine the same function $\mathbb{Z}_p \to \mathbb{Z}_p$, namely, the identity map $a \mapsto a$, because Fermat's theorem says that $a^p \equiv a$ for all $a \in \mathbb{Z}$; that is, $[a^p] = [a]$. Therefore, $x^p - x \in \ker \varphi$. This example gives one reason why our definition of polynomials is so formal.

One can generalize the example just given. If F is any field, then there are always infinitely many polynomials over F; indeed, if $m \neq n$, then

$x^m \neq x^n$. But if F is any *finite* field (and we shall see that there are finite fields other than $\mathbb{Z}_p$), then there are only finitely many functions $F \to F$, and hence only finitely many polynomial functions: $P(F)$ is finite. In this case, there can be no bijection between $F[x]$ and $P(F)$, for one is infinite and the other is finite. The next result shows that this pathology vanishes when the coefficient field F is infinite: polynomials over an infinite field F and polynomials functions over F are essentially the same.

Corollary 23. *If F is a field and $f(x) \in F[x]$, denote its polynomial function $F \to F$ by f. If F is infinite, then the function $\varphi : F[x] \to P(F)$, given by $\varphi : f(x) \mapsto f$, is an isomorphism.*

Proof. It suffices to prove that $\ker \varphi = \{0\}$. Suppose that $f(x) \in \ker \varphi$ is not the zero polynomial, and let $n = \partial(f)$. Since $f(a) = 0$ for all $a \in F$, each of the infinitely many elements $a \in F$ is a root of $f(x)$, and this contradicts Theorem 22. ●

Exercises

40. Prove that there are domains R containing a pair of elements having no gcd. (Hint. Let F be a field and let R be the subring of $F[x]$ consisting of all polynomials having no linear term; i.e., $f(x) \in R$ if and only if

$$f(x) = a_0 + a_2 x^2 + a_3 x^3 + \cdots .$$

Show that x^5 and x^6 have no gcd by noting that their monic divisors are $1, x^2$, and x^3, none of which is divisible in R by the other two.)

41. (i) Define the gcd of integers $a_1, \dots, a_n$ to be a positive integer d which is a *common divisor*, i.e., $d \mid a_i$ for all i, that is divisible by every common divisor. Prove that the gcd d of $a_1, \dots, a_n$ exists, and that d is a linear combination of $a_1, \dots, a_n$. (Hint. Let d be the positive generator of the ideal in $\mathbb{Z}$ generated by $a_1, \dots, a_n$.)

(ii) Define the gcd of polynomials $f_1, \dots, f_n \in F[x]$, where F is a field, to be a monic polynomial d which is a *common divisor*, i.e., $d \mid f_i$ for all i, that is divisible by every common divisor. Prove the generalization of Corollary 16 that the gcd d of $f_1, \dots, f_n$ exists, and that d is a linear combination of $f_1, \dots, f_n$.

42. Prove that if $a_1, a_2, \dots, a_n$ are distinct elements in a field F, then for all i, the polynomials $x - a_{i+1}$ and $(x - a_1)(x - a_2) \cdots (x - a_i)$ are relatively prime.

43. In the ring $R = \mathbb{Z}[x]$, show that x and 2 are relatively prime, but there are no polynomials $f(x)$ and $g(x) \in \mathbb{Z}[x]$ with $1 = xf(x) + 2g(x)$.

44. Let $f(x) = \prod(x-a_i) \in F[x]$, where F is a field and $a_i \in F$ for all i. Show that $f(x)$ has **no repeated roots** [i.e., $f(x)$ is not a multiple of $(x - a)^2$ for any $a \in F$] if and only if $(f(x), f'(x)) = 1$, where $f'(x)$ is the derivative of $f(x)$.

45. Find the gcd of $x^3 - 2x^2 + 1$ and $x^2 - x - 3$ in $\mathbb{Q}[x]$ and express it as a linear combination.

46. Prove that $\mathbb{Z}_2[x]/I$ is a field, where $p(x) = x^3 + x + 1 \in \mathbb{Z}_2[x]$ and $I = (p(x))$.

47. If R is a ring and $a \in R$, let $e_a : R[x] \to R$ be evaluation at a. Prove that $\ker e_a$ consists of all the polynomials over R having a as a root, and so $\ker e_a = (x - a)$, the principal ideal generated by $x - a$.

48. Let F be a field, and let $f(x), g(x) \in F[x]$. Prove that if $\partial(f) \le \partial(g) = n$ and if $f(a) = g(a)$ for $n + 1$ elements $a \in F$, then $f(x) = g(x)$.

Prime Ideals and Maximal Ideals

The notion of prime number can be generalized to polynomials.

Definition. Let F be a field. A nonzero polynomial $p(x) \in F[x]$ is *irreducible[4] over* F if $\partial(p) \ge 1$ and there is no factorization $p(x) = f(x)g(x)$ in $F[x]$ with $\partial(f) < \partial(p)$ and $\partial(g) < \partial(p)$.

Notice that irreducibility does depend on the coefficient field F. For example, $x^2 + 1$ is irreducible over $\mathbb{R}$, but it factors over $\mathbb{C}$. It is easy to see that linear polynomials (degree 1) are irreducible over any field F for which they are defined. It follows from Corollary 21 that irreducible polynomials of degree ≥ 2 over a field F have no roots in F. The converse is false, however, for $f(x) = x^4 + 2x^2 + 1 = (x^2 + 1)^2$ factors over $\mathbb{R}$, but it has no real roots.

---

[4]This notion can be generalized to any ring R. A nonzero element $r \in R$ is called *irreducible* if r is not a unit and, in every factorization $r = st$ in R, either s or t is a unit. If F is a field and $R = F[x]$, then this notion coincides with our definition of irreducible polynomial. In $\mathbb{Z}[x]$, however, $2x + 2 = 2(x + 1)$ is not irreducible, yet it does not factor into polynomials each of which has smaller degree.

Let $p(x)$, $f(x) \in F[x]$, where F is a field. If $p(x)$ is a monic irreducible polynomial, then its only monic divisors are 1 and $p(x)$; hence, the gcd (p, f) is either 1 or $p(x)$. It follows that if $p(x)$ does not divide $f(x)$, then $p(x)$ and $f(x)$ are relatively prime.

Definition. An ideal I in a ring R is called a *prime ideal* if it is a proper ideal and $ab \in I$ implies $a \in I$ or $b \in I$.

Example 14. We claim that if $p \geq 2$, then the ideal (p) in $\mathbb{Z}$ is a prime ideal if and only if p is a prime. If p is prime and $ab \in (p)$, then $p \mid ab$. By Euclid's lemma, either $p \mid a$ or $p \mid b$; that is, either $a \in (p)$ or $b \in (p)$. Therefore, (p) is a prime ideal.

Conversely, if p is not a prime, then it has a factorization $p = ab$ with $a < p$ and $b < p$. It follows that neither a nor b lies in (p), and so (p) is not a prime ideal.

Theorem 24. *If F is a field, then a nonzero polynomial $p(x) \in F[x]$ is irreducible if and only if $(p(x))$ is a prime ideal.*

Proof. Suppose that $p(x)$ is irreducible. If $ab \in (p)$, then $p \mid ab$, and so Euclid's lemma gives either $p \mid a$ or $p \mid b$. Thus, $a \in (p)$ or $b \in (p)$. Finally, (p) is a proper ideal; otherwise, $1 \in R = (p)$, and so there is a polynomial $f(x)$ with $1 = p(x)f(x)$. But the constant 1 has degree 0, whereas

$$\partial(pf) = \partial(p) + \partial(f) \geq \partial(p) \geq 1.$$

This contradiction shows that (p) is a proper ideal, and hence it is a prime ideal.

Conversely, suppose that $p(x)$ is not irreducible; there is thus a factorization

$$p(x) = a(x)b(x)$$

with $\partial(a) < \partial(p)$ and $\partial(b) < \partial(p)$. As every nonzero polynomial in (p) has degree $\geq \partial(p)$, it follows that neither a nor b lies in (p), and so (p) is not a prime ideal. •

Theorem 25. *A proper ideal I in R is a prime ideal if and only if R/I is a domain.*

Proof. Let I be a prime ideal. If $0 = (a+I)(b+I) = ab+I$, then $ab \in I$. Since I is a prime ideal, either $a \in I$ or $b \in I$; that is, either $a + I = 0$ or $b + I = 0$. Hence, R/I is a domain. The converse is just as easy. •

Definition. An ideal I in a ring R is a *maximal ideal* if it is a proper ideal and there is no ideal J with $I \subsetneqq J \subsetneqq R$.

Theorem 26. *A proper ideal I in a ring R is a maximal ideal if and only if R/I is a field.*

Proof. The Correspondence Theorem (Exercise 38) shows that I is a maximal ideal if and only if R/I has no ideals other than $\{0\}$ and R/I itself; Exercise 33 shows that this property holds if and only if R/I is a field. •

Corollary 27. *Every maximal ideal I in a ring R is a prime ideal.*

Proof. If I is a maximal ideal, then R/I is a field. Since every field is a domain, R/I is a domain, and so I is a prime ideal. •

The converse of the last corollary is false. For example, the principal ideal (x) in $\mathbb{Z}[x]$ is prime but not maximal; by Exercise 37, we have

$$\mathbb{Z}[x]/(x) \cong \mathbb{Z},$$

and $\mathbb{Z}$ is a domain but not a field.

Theorem 28. *If R is a principal ideal domain, then every nonzero prime ideal I is a maximal ideal.*

Proof. Assume there is an ideal $J \neq I$ with $I \subset J \subset R$. Since R is a PID, $I = (a)$ and $J = (b)$ for some $a, b \in R$. Now $a \in J$ implies that $a = rb$ for some $r \in R$, and so $rb \in I$. Since I is prime, either $r \in I$ or $b \in I$. If $b \in I$, then $J \subset I$, a contradiction. If $r \in I$, then $r = sa$ for some $s \in R$, and so $a = rb = sab$; hence $1 = sb$ and $J = (b) = R$, by Exercise 32(i). Therefore, I is maximal. •

Corollary 29. *If F is a field and $p(x) \in F[x]$ is irreducible, then the quotient ring $F[x]/(p(x))$ is a field containing (an isomorphic copy of) F and a root of $p(x)$.*

Proof. Since $p(x)$ is irreducible, the principal ideal $I = (p(x))$ is a nonzero prime ideal; since $F[x]$ is a PID, I is a maximal ideal, and so $E = F[x]/I$ is a field. Now the map $a \mapsto a + I$ is an isomorphism from F to $F' = \{a + I : a \in F\} \subset E$ (one usually identifies F with this subfield F' of E).

Let $\theta = x + I \in E$; we claim that θ is a root of $p(x)$. Write $p(x) = a_0 + a_1 x + \cdots + a_n x^n$, where $a_i \in F$. Then, in E:

$$
\begin{aligned}
p(\theta) &= (a_0 + I) + (a_1 + I)\theta + \cdots + (a_n + I)\theta^n \\
&= (a_0 + I) + (a_1 + I)(x + I) + \cdots + (a_n + I)(x + I)^n \\
&= (a_0 + I) + (a_1 x + I) + \cdots + (a_n x^n + I) \\
&= a_0 + a_1 x + \cdots + a_n x^n + I \\
&= p(x) + I = I,
\end{aligned}
$$

because $I = (p(x))$. But $I = 0 + I$ is the zero element of $F[x]/I$, and hence θ is a root of $p(x)$. •

For example, $x^2 + 1$ is an irreducible polynomial in $\mathbb{R}[x]$, and the quotient ring $\mathbb{R}[x]/(x^2 + 1)$ is a field containing $\mathbb{R}$ and an element i with $i^2 = -1$, namely, $i = x + I$. We have seen, in Example 10, that $\mathbb{R}[x]/(x^2 + 1)$ is isomorphic to the field of complex numbers $\mathbb{C}$.

Definition. A polynomial $f(x) \in F[x]$ **splits over** F if it is a product of linear factors in $F[x]$.

Of course, $f(x)$ splits over F if and only if F contains all the roots of $f(x)$.

Theorem 30 (Kronecker). *Let $f(x) \in F[x]$, where F is a field. There exists a field E containing F over which $f(x)$ splits.*

Proof. The proof is by induction on $\partial(f)$. If $\partial(f) = 1$, then $f(x)$ is linear and we can choose $E = F$. If $\partial(f) > 1$, write $f(x) = p(x)g(x)$, where $p(x)$ is irreducible. The corollary provides a field B containing F and a root θ of $p(x)$. Hence $p(x) = (x - \theta)h(x)$ in $B[x]$. By induction, there is a field E containing B over which $h(x)g(x)$, hence $f(x)$, splits. •

We now modify, for a polynomial $f(x) \in F[x]$, the definition of repeated roots appearing in Exercise 44 so that roots may belong to some larger field of coefficients than F.

Definition. If F is a field and $f(x) \in F[x]$, then $f(x)$ has **repeated roots** if there is a field E containing F and a factorization in $E[x]$ of the form

$$
f(x) = (x - a)^2 h(x).
$$

Using Exercise 44 and Corollary 18, one sees that $f(x)$ has no repeated roots if and only if $(f, f') = 1$, where $f'(x)$ is the derivative of $f(x)$. Notice that the criterion $(f, f') = 1$ can be checked over the original field F; there is no need to examine factorizations over E.

Kronecker's theorem can be used to construct finite fields other than the fields $\mathbb{Z}_p$. Before giving the construction, we introduce an important property of fields.

Definition. The *prime field* of a field F is the intersection of all the subfields of F.

By Exercise 19, the prime field is a subfield.

Theorem 31. *If F is a field, then its prime field is isomorphic to either $\mathbb{Q}$ or $\mathbb{Z}_p$ for some prime p.*

Proof. Define $\chi : \mathbb{Z} \to F$ by $n \mapsto n1$ (where 1 is the "one" in F); it is easy to see that χ is a ring map. If $I = \ker \chi$, then $\mathbb{Z}/I$ is a domain (because it is isomorphic to a subring of the field F). Therefore, I is a prime ideal, and hence $I = (0)$ or $I = (p)$ for some prime p. If $I = (0)$, then χ imbeds $\mathbb{Z}$ in F. By Exercise 25, the prime field is isomorphic to $\mathbb{Q}$ in this case. If $I = (p)$, the first isomorphism theorem gives im $\chi \cong \mathbb{Z}/(p) = \mathbb{Z}_p$, which is a field; hence im χ is the prime field of F. •

Definition. A field has *characteristic 0* if its prime field is isomorphic to $\mathbb{Q}$; it has *characteristic p* if its prime field is isomorphic to $\mathbb{Z}_p$.

Lemma 32. *Let F be a field of characteristic $p > 0$.*

(i) *For all $a \in F$, we have $pa = 0$.*

(ii) *$(a + b)^p = a^p + b^p$ for all $a, b \in F$.*

(iii) *$(a + b)^{p^k} = a^{p^k} + b^{p^k}$ for all $a, b \in F$ and all $k \geq 1$.*

Proof. (i) For the moment, let us denote "one" in F by e. Now pa means the sum of p terms each equal to a:

$$pa = a + \cdots + a = (e + \cdots + e)a.$$

In $\mathbb{Z}_p$, however, the sum of p terms each equal to [1] is 0. Since F has characteristic p, we have $e + \cdots + e = 0$, and so $pa = 0$ in F.

(ii) The binomial theorem gives

$$(a+b)^p = a^p + \sum_{i=1}^{p-1} \binom{p}{i} a^i b^{p-i} + b^p.$$

By Exercise 7, $\binom{p}{i} = 0$ in $\mathbb{Z}_p$ for all $1 \le i \le p-1$.

(iii) The proof is by induction on $k \ge 1$, the base step being part (ii). For the inductive step,

$$(a+b)^{p^{k+1}} = [(a+b)^{p^k}]^p = [a^{p^k} + b^{p^k}]^p = a^{p^{k+1}} + b^{p^{k+1}}. \quad \bullet$$

It follows from this lemma that if F is a field of characteristic p and if $q = p^k$, then the function $a \mapsto a^q$ is a ring homomorphism from F to itself.

The following elementary remark is very useful. If F is a subfield of a field E, then the additive group of E may be viewed as a vector space over F. Define scalar multiplication by letting $c\alpha$, for $c \in F$ and $\alpha \in E$, be the product of the two elements c and α under the given multiplication on E. Viewing the appropriate axioms in the definition of a field in this light, one can see that they are also the axioms of a vector space over F. In particular, a finite field E must have characteristic p for some prime p, for its prime field cannot be $\mathbb{Q}$, and so it is a vector space over $\mathbb{Z}_p$. If $\{\alpha_1, \dots, \alpha_n\}$ is an ordered basis of E, then each $a \in E$ has coordinates $(c_1, \dots, c_n)$ for c_i in $\mathbb{Z}_p$. Therefore, every finite field has p^n elements, for some prime p and some positive integer n.

Theorem 33 (Galois). *For every prime p and every positive integer n, there exists a field having exactly p^n elements.*

Proof. If there were a field K with $|K| = p^n = q$, then $K^\# = K - \{0\}$ would be a multiplicative group of order $q - 1$; by Lagrange's theorem, $a^{q-1} = 1$ for all $a \in K^\#$. It follows that every element of K would be a root of the polynomial

$$g(x) = x^q - x.$$

We now begin the construction. By Kronecker's theorem, there is a field E containing $\mathbb{Z}_p$ over which $g(x)$ splits. Define $F = \{\alpha \in E : g(\alpha) = 0\}$; that is, F is the set of all the roots of $g(x)$. Since the derivative $g'(x) = qx^{q-1} - 1 = -1$ (because $q = p^n$ and E has characteristic p), Lemma 32 shows that the gcd $(g, g') = 1$, and so $g(x)$ has no repeated roots; that is, $|F| = q = p^n$.

We claim that F is a field, which will complete the proof. If $a, b \in F$, then $a^q = a$ and $b^q = b$. Therefore, $(ab)^q = a^q b^q = ab$, and $ab \in F$. By Lemma 32(iii), replacing b by $-b$, we have $(a - b)^q = a^q - b^q = a - b$, so that $a - b \in F$. Finally, if $a \neq 0$, then $a^{q-1} = 1$ so that $a^{-1} = a^{q-2} \in F$ (because F is closed under multiplication).   •

In Corollary 53 we shall see that any two fields of order p^n are isomorphic. It will follow that there are no finite fields other than those just constructed.

Exercises

49. A polynomial $p(x) \in F[x]$ of degree 2 or 3 is irreducible over F if and only if F contains no root of $p(x)$. (This is false for degree 4: the polynomial $(x^2 + 1)^2$ factors in $\mathbb{R}[x]$, but it has no real roots.)

50. Let $p(x) \in F[x]$ be irreducible. If $g(x) \in F[x]$ is not constant, then either $(p(x), g(x)) = 1$ or $p(x) \mid g(x)$.

51. (i) Every nonzero polynomial $f(x)$ in $F[x]$ has a factorization of the form
$$f(x) = ap_1(x) \cdots p_t(x),$$
where a is a nonzero constant and the $p_i(x)$ are (not necessarily distinct) monic irreducible polynomials;

(ii) the factors and their multiplicities in this factorization are uniquely determined.

(This analogue of the fundamental theorem of arithmetic has the same proof as that theorem: if also $f(x) = bq_1(x) \dots q_s(x)$, where b is constant and the $q_j(x)$ are monic and irreducible, then uniqueness is proved by Euclid's lemma and induction on $\max\{t, s\}$. One calls $F[x]$ a *unique factorization domain* when one wishes to call attention to this property of it.)

52. Let $f(x) = ap_1(x)^{k_1} \cdots p_t(x)^{k_t}$ and $g(x) = bp_1(x)^{n_1} \cdots p_t(x)^{n_t}$, where $k_i \geq 0, n_i \geq 0, a, b$ are nonzero constants, and the $p_i(x)$ are distinct monic irreducible polynomials (zero exponents allow one to have the same $p_i(x)$ in both factorizations). Prove that
$$\gcd(f, g) = p_1(x)^{m_1} \cdots p_t(x)^{m_t}$$
and
$$\mathrm{lcm}(f, g) = p_1(x)^{M_1} \cdots p_t(x)^{M_t},$$
where $m_i = \min\{k_i, n_i\}$ and $M_i = \max\{k_i, n_i\}$.

53. (i) Prove that the zero ideal in a ring R is a prime ideal if and only if R is a domain.

    (ii) Prove that the zero ideal in a ring R is a maximal ideal if and only if R is a field.

54. The ideal I in $\mathbb{Z}[x]$ consisting of all polynomials having even constant term is a maximal ideal.

55. Let $f(x), g(x) \in F[x]$. Then $(f, g) \neq 1$ if and only if there is a field E containing both F and a common root of $f(x)$ and $g(x)$.

56. (i) Prove that if $f(x) \in \mathbb{Z}_p[x]$, then $(f(x))^p = f(x^p)$. (Hint: Use Fermat's theorem: $a^p \equiv a \bmod p$.)

    (ii) Show that the first part of this exercise may be false if $\mathbb{Z}_p$ is replaced by an infinite field of characteristic p.

57. Exhibit an infinite field of characteristic p. (Hint: Exercise 20.)

58. If F is a field, prove that the kernel of any evaluation map $F[x] \to F$ is a maximal ideal.

59. If F is a field of characteristic 0 and $p(x) \in F[x]$ is irreducible, then $p(x)$ has no repeated roots. (Hint: Consider $(p(x), p'(x))$.)

60. Use Kronecker's theorem to construct a field with four elements by adjoining a suitable root of $x^4 - x$ to $\mathbb{Z}_2$.

61. Give the addition and multiplication tables of a field having eight elements. (Hint: Factor $x^8 - x$ over $\mathbb{Z}_2$.)

62. Show that a field with four elements is not (isomorphic to) a subfield of a field with eight elements.

Irreducible Polynomials

Our next project is to find some criteria for irreducibility of polynomials; this is usually difficult, and it is unsolved in general.

We begin with an elementary result, using Exercise 29: If $\sigma : R \to S$ is a ring map, then $\sigma^* : R[x] \to S[x]$, defined by

$$\sigma^* : \sum r_i x^i \mapsto \sum \sigma(r_i) x^i,$$

is also a map of rings.

Theorem 34. *Let R be a domain and F be a field, let $\sigma : R \to F$ be a ring map, and let $p(x) \in R[x]$. If $\partial(\sigma^*(p)) = \partial(p)$ and if $\sigma^*(p(x))$ is irreducible in $F[x]$, then $p(x)$ is not a product of two polynomials in $R[x]$ each of degree $< \partial(p)$.*

Remark. Note that the degree condition is satisfied if $p(x)$ is monic.

Proof. Suppose that $p(x) = f(x)g(x)$ in $R[x]$ with $\partial(f) < \partial(p)$ and $\partial(g) < \partial(p)$. In $F[x]$, we have $\sigma^*(p) = \sigma^*(f)\sigma^*(g)$. Since $\sigma^*(p)$ is irreducible, we may assume that $\partial(\sigma^*(f)) = 0$. But

$$
\begin{aligned}
\partial(p) &= \partial(\sigma^*(p)) \\
&= \partial(\sigma^*(f)) + \partial(\sigma^*(g)) \\
&= \partial(\sigma^*(g)) \\
&\leq \partial(g) \\
&< \partial(p).
\end{aligned}
$$

This contradiction completes the proof.   •

Example 15. Consider $f(x) = 8x^3 - 6x - 1$ in $\mathbb{Z}[x]$. We will use the theorem by making a suitable choice of prime p and taking $\sigma : \mathbb{Z} \to \mathbb{Z}_p$ to be the natural map; thus, σ^* reduces the coefficients of $f(x)$ mod p. If we choose $p = 2$, then the degree condition is not satisfied because $\sigma^*(f)$ has degree 0. If $p = 3$, then

$$
\sigma^*(f) = -x^3 - 1 = -(x+1)(x^2 - x + 1),
$$

and $\sigma^*(f)$ is not irreducible. If $p = 5$, then

$$
\sigma^*(f) = 3x^3 - x - 1;
$$

this is irreducible, by Exercise 49, for it has no roots in $\mathbb{Z}_5$. It follows from Theorem 34 that $f(x)$ is not a product in $\mathbb{Z}[x]$ of polynomials of lower degree.

Theorem 34 does not always apply. We shall see, in Exercise 67, that $f(x) = x^4 - 10x^2 + 1$ is irreducible in $\mathbb{Q}[x]$; in Example 26, we will show that $f(x)$ factors mod p for every prime p.

We now have a way to see whether certain polynomials in $\mathbb{Z}[x]$ factor into polynomials of smaller degree, but we are really interested in whether polynomials are irreducible in $\mathbb{Q}[x]$. A result of Gauss will solve this problem.

Definition. A polynomial $f(x) = a_0 + a_1 x + \cdots + a_n x^n \in \mathbb{Z}[x]$ is called *primitive* if the gcd[5] of its coefficients is 1.

If d is the gcd of the coefficients of $f(x)$, then $(1/d)f(x)$ is a primitive polynomial.

Observe that if $f(x)$ is not primitive, then there exists a prime p which divides each of its coefficients: if the gcd of the coefficients is d, let p be any prime divisor of d.

Lemma 35 (Gauss's Lemma). *The product of two primitive polynomials $f(x)$ and $g(x)$ is itself primitive.*

Proof.[6] Assume that the product $f(x)g(x)$ is not primitive, so there is some prime p dividing each of its coefficients. Let $\sigma : \mathbb{Z} \to \mathbb{Z}_p$ be the natural map, and consider the ring map $\sigma^* : \mathbb{Z}[x] \to \mathbb{Z}_p[x]$ reducing coefficients mod p. Now

$$\sigma^*(fg) = \sigma^*(f)\sigma^*(g).$$

But $\sigma^*(fg) = 0$ in $\mathbb{Z}_p[x]$ while $\sigma^*(f) \neq 0$ and $\sigma^*(g) \neq 0$, and this contradicts the fact that $\mathbb{Z}_p[x]$ is a domain.   •

Lemma 36. *Every nonzero $f(x) \in \mathbb{Q}[x]$ has a unique factorization*

$$f(x) = c(f)f^*(x),$$

where $c(f) \in \mathbb{Q}$ is positive and $f^*(x) \in \mathbb{Z}[x]$ is primitive.

Remark. The positive rational $c(f)$ is called the **content** of $f(x)$.

Proof. Let $f(x) = (a_0/b_0) + (a_1/b_1)x + \cdots + (a_n/b_n)x^n \in \mathbb{Q}[x]$. Define $B = b_0 \cdots b_n$, so that $g(x) = Bf(x) \in \mathbb{Z}[x]$. Now define $D = \pm d$, where d is the gcd of the coefficients of $g(x)$; the sign is chosen to make D/B positive. Now $(B/D)f(x) = (1/D)g(x)$ lies in $\mathbb{Z}[x]$, and it is a primitive polynomial. If we define $c(f) = D/B$ and $f^*(x) = (B/D)f(x)$, then $f(x) = c(f)f^*(x)$ is a desired factorization.

Suppose that $f(x) = eh(x)$ is a second such factorization, so that e is a positive rational and $h(x) \in \mathbb{Z}[x]$ is primitive. Now $c(f)f^*(x) = f(x) = eh(x)$, so that $f^*(x) = [(e/c(f)]h(x)$. Write $e/c(f)$ in lowest terms: $e/c(f) = u/v$, where u and v are relatively prime positive integers. The equation $vf^*(x) = uh(x)$ holds in $\mathbb{Z}[x]$; equating like coefficients, v is

---

[5]Exercise 41 defines the gcd of finitely many integers and shows that it always exists.

[6]This elegant proof of Gauss's lemma was shown me by Peter Cameron.

a common divisor of each coefficient of $uh(x)$. Since $(u, v) = 1$, Euclid's lemma in $\mathbb{Z}$ shows that v is a (positive) common divisor of each coefficient of $h(x)$. Since $h(x)$ is primitive, it follows that $v = 1$. A similar argument shows that $u = 1$. Therefore, $e/c(f) = u/v = 1$, so that $e = c(f)$ and hence $f^*(x) = h(x)$.   •

Corollary 37. *If $f(x) \in \mathbb{Z}[x]$, then $c(f) \in \mathbb{Z}$.*

Proof. If d is the gcd of the coefficients of $f(x)$, then $(1/d)f(x) \in \mathbb{Z}[x]$ is primitive. Since $f(x) = d[(1/d)f(x)]$ is a product of a positive rational d (even an integer) and a primitive polynomial, the uniqueness in the lemma gives $c(f) = d \in \mathbb{Z}$.   •

Corollary 38. *If $f(x) \in \mathbb{Q}[x]$ factors as $f(x) = g(x)h(x)$ in $\mathbb{Q}[x]$, then*

$$c(f) = c(g)c(h) \text{ and } f^*(x) = g^*(x)h^*(x).$$

Proof. We have

$$\begin{aligned} f(x) &= g(x)h(x) \\ &= [c(g)g^*(x)][c(h)h^*(x)] \\ &= c(g)c(h)g^*(x)h^*(x). \end{aligned}$$

Since $c(g)c(h)$ is a positive rational, and since the product of two primitive polynomials is primitive, the uniqueness of the factorization in the preceding lemma gives $c(f) = c(g)c(h)$ and $f^*(x) = g^*(x)h^*(x)$.   •

Theorem 39   (Gauss). *If $p(x) \in \mathbb{Z}[x]$ is not a product of two polynomials in $\mathbb{Z}[x]$ each of degree $< \partial(p)$, then $p(x)$ is irreducible in $\mathbb{Q}[x]$.*

Proof. If $f(x) = g(x)h(x)$ in $\mathbb{Q}[x]$, then $f(x) = c(g)c(h)g^*(x)h^*(x)$ in $\mathbb{Q}[x]$, where g^* and h^* are primitive polynomials in $\mathbb{Z}[x]$. But $c(g)c(h) = c(f) \in \mathbb{Z}$, by Corollary 37. Therefore, $f(x) = [c(f)g^*(x)]h^*(x)$ is a factorization in $\mathbb{Z}[x]$.   •

Remark. The proof of this last theorem can be adapted to a more general setting: replace $\mathbb{Z}$ and $\mathbb{Q}$ by a unique factorization domain and its field of fractions. This is the main ingredient of the proof that if R is a unique factorization domain, then so is $R[x]$; it follows that if F is a field, then $F[x_1, \ldots, x_n]$ is a unique factorization domain.

Theorem 40   (Eisenstein Criterion). *Let $f(x) = a_0 + a_1 x + \cdots + a_n x^n \in \mathbb{Z}[x]$. If there is a prime p dividing a_i for all $i < n$, but with p not dividing a_n and p^2 not dividing a_0, then $f(x)$ is irreducible in $\mathbb{Q}[x]$.*

Proof. Let

$$f(x) = g(x)h(x) = (b_0 + b_1x + \cdots + b_mx^m)(c_0 + c_1x + \cdots + c_kx^k);$$

by Theorem 39, we may assume that both g and h lie in $\mathbb{Z}[x]$. By hypothesis, $p \mid a_0 = b_0c_0$ so that $p \mid b_0$ or $p \mid c_0$, by Euclid's lemma in $\mathbb{Z}$; since p^2 does not divide a_0, only one of them is divisible by p, say, $p \mid c_0$ but p does not divide b_0. The leading coefficient $a_n = b_mc_k$ is not divisible by p, so that p does not divide c_k (or b_m). Let c_r be the first coefficient not divisible by p (so p does divide $c_0, \ldots, c_{r-1}$). If $r < n$, then $p \mid a_r$, and $b_0c_r = a_r - (b_1c_{r-1} + \cdots + b_rc_0)$ is divisible by p; hence $p \mid b_0c_r$, contradicting Euclid's lemma (because p divides neither factor). It follows that $r = n$, hence $k = 0$, and $h(x)$ is constant. Therefore, $f(x)$ is irreducible.   •

Remark. The following more elegant proof of the Eisenstein criterion is due to Peter Cameron. Suppose that $f(x) = g(x)h(x)$ in $\mathbb{Z}[x]$, where $g(x) = b_0 + b_1x + \cdots + b_mx^m$ has $m < n$ and $h(x) = c_0 + c_1x + \cdots + c_kx^k$ has $k < n$. Consider the map $\sigma^* : \mathbb{Z}[x] \to \mathbb{Z}_p[x]$ reducing coefficients mod p. In $\mathbb{Z}_p[x]$, we have $\sigma^*(f) = a_nx^n$; thus, $\sigma^*(f)$ is a constant times monic irreducibles all equal to x. But $\sigma^*(f) = \sigma^*(g)\sigma^*(h)$, so that unique factorization (see Exercise 51) shows that $\sigma^*(g)$ and $\sigma^*(h)$ have similar factorizations. Therefore, $\sigma^*(g) = b_mx^m$ and $p \mid b_i$ for all $i < m$; similarly, $\sigma^*(h) = c_kx^k$ and $p \mid c_j$ for all $j < k$. In particular, $p \mid b_0$ and $p \mid c_0$ and so $p^2 \mid b_0c_0 = a_0$, a contradiction.

The Eisenstein criterion shows that $x^5 - 4x + 2$ is irreducible over $\mathbb{Q}$; this polynomial does not surrender easily to our first criterion.

Definition. If p is a prime, then the pth *cyclotomic polynomial* is

$$\Phi_p(x) = (x^p - 1)/(x - 1) = x^{p-1} + x^{p-2} + \cdots + x + 1.$$

Corollary 41.   *The pth cyclotomic polynomial $\Phi_p(x)$ is irreducible in $\mathbb{Q}[x]$ for every prime p.*

Proof. Recall Exercise 66: A polynomial $f(x)$ is irreducible if and only if $f(x + c)$ is irreducible, where c is a constant. In particular,

$$\Phi_p(x) = (x^p - 1)/(x - 1)$$

is irreducible if and only if

$$\Phi_p(x + 1) = ((x + 1)^p - 1)/x$$

is irreducible. The latter polynomial is $x^{p-1} + px^{p-2} + \binom{p}{2}x^{p-3} + \cdots +$ p, where $\binom{p}{i}$ is the binomial coefficient. Since p is prime, Exercise 7 shows that Eisenstein's criterion applies; we conclude that $\Phi_p(x)$ is irreducible.   •

If n is not prime, then $x^{n-1} + x^{n-2} + \cdots + x + 1$ factors in $\mathbb{Q}[x]$. For example,

$$x^3 + x^2 + x + 1 = (x + 1)(x^2 + 1).$$

Corollary 42.   *If an integer a is not a perfect square, then $x^n - a$ is irreducible in $\mathbb{Q}[x]$ for every $n \geq 2$.*

Proof. Since $a \neq \pm 1$, there is some prime p dividing a, and Eisenstein's criterion applies with this prime.   •

This last corollary shows that there are irreducible polynomials over $\mathbb{Q}$ of arbitrary degree n.

Exercises

63. Let $f(x) = a_0 + a_1x + \cdots + a_nx^n \in \mathbb{Z}[x]$. If r/s is a rational root of $f(x)$, where r/s is in lowest terms, i.e., $(r, s) = 1$, then $r \mid a_0$ and $s \mid a_n$. Conclude that any rational root of a monic polynomial in $\mathbb{Z}[x]$ must be an integer.

64. Test whether the following polynomials factor in $\mathbb{Q}[x]$:
    (i) $3x^2 - 7x - 5$;
    (ii) $6x^3 - 3x - 18$;
    (iii) $x^3 - 7x + 1$.
    (iv) $x^3 - 9x - 9$.

65. Let F be a field. Prove that if $a_0 + a_1x + \cdots + a_nx^n \in F[x]$ is irreducible, then so is $a_n + a_{n-1}x + \cdots + a_0x^n$.

66. If $c \in R$, where R is a ring, then the map $f(x) \mapsto f(x + c)$ is an isomorphism of the ring $R[x]$ with itself. Conclude, when R is a field, that $p(x)$ is irreducible if and only if $p(x + c)$ is irreducible.

67. Prove that $f(x) = x^4 - 10x^2 + 1$ is irreducible in $\mathbb{Q}[x]$. (Hint. Use Exercise 63 to show that $f(x)$ has no rational roots; then show that there are no rationals a, b and c with

$$x^4 - 10x^2 + 1 = (x^2 + ax + b)(x^2 - ax + c).)$$

Classical Formulas

We now derive the classical formulas for the roots of quadratics, cubics, and quartics.

Definition. A polynomial $f(x)$ of degree n is called **reduced** if it has no x^{n-1} term; that is,

$$f(x) = r_n x^n + r_{n-2} x^{n-2} + r_{n-3} x^{n-3} + \cdots$$

Lemma 43. *If $f(X) = a_n X^n + a_{n-1} X^{n-1} + a_{n-2} X^{n-2} + \cdots$, then replacing X by $x - a_{n-1}/n$ gives a reduced polynomial*

$$\widetilde{f}(x) = f(x - a_{n-1}/n);$$

moreover, if u is a root of $\widetilde{f}(x)$, then $u - a_{n-1}/n$ is a root of $f(X)$.

Proof. The first statement is a straightforward calculation, and the second statement follows from the equation $0 = \widetilde{f}(u) = f(u - a_{n-1}/n)$.   •

The quadratic formula is usually proved by "completing the square," but we shall do it in a way that anticipates the derivations of its generalizations to cubics and quartics. Consider the quadratic

$$X^2 + bX + c.$$

The substitution replacing X by $x - \frac{1}{2}b$ gives the reduced quadratic

$$x^2 + c - \tfrac{1}{4}b^2$$

having roots $u = \pm\frac{1}{2}\sqrt{b^2 - 4c}$. By Lemma 43, one obtains the familiar formula for the roots of the original quadratic:

$$-\tfrac{1}{2}b \pm \tfrac{1}{2}\sqrt{b^2 - 4c}.$$

Before discussing formulas for the roots of a polynomial $f(x)$ of higher degree, we must say that if $f(x) \in \mathbb{Z}[x]$, one should first use Exercise 63 to see if it has any rational roots. If u is a root of a cubic polynomial $f(x)$, for example, then its remaining roots are the roots of the quadratic polynomial $f(x)/(x - u)$.

The reduced polynomial arising from a cubic $X^3 + aX^2 + bX + c$ has the form

$$g(x) = x^3 + qx + r;$$

by Lemma 43, a formula for the roots of $g(x)$ will give a formula for the roots of the original one. The coming formula is essentially due to Scipio del Ferro (ca. 1515); a similar formula was discovered by Tartaglia about the same time, and both appeared in print for the first time in the book of Cardan (1545).

Let u be a root of $g(x)$, and choose numbers y and z with $u = y + z$. Then

$$u^3 = (y+z)^3 = y^3 + z^3 + 3(y^2z + yz^2) = y^3 + z^3 + 3uyz.$$

Therefore,

(1)
$$y^3 + z^3 + (3yz + q)u + r = 0.$$

So far we have imposed only one constraint on y and z, namely, $u = y + z$. By Exercise 68, we may impose a second constraint:

(2)
$$yz = -q/3,$$

so that, in Eq. (1), the linear term in u vanishes. We now have

$$y^3 + z^3 = -r$$

and

$$y^3z^3 = -q^3/27.$$

These two equations can be solved for y^3 and z^3. In detail,

$$y^3 - q^3/27y^3 = -r,$$

and hence

$$y^6 + ry^3 - q^3/27 = 0.$$

The quadratic formula gives

(3)
$$y^3 = \tfrac{1}{2}\left(-r + \sqrt{r^2 + 4q^3/27}\right),$$

and Eq. (2) gives $z = -q/3y$. Having found one root $u = y + z$ of $g(x)$, one can find the other two as the roots of the quadratic $g(x)/(x - u)$.

Here is an explicit formula for the other two roots, in contrast to the method just described for finding them. If $\omega = e^{2\pi i/3}$ is a cube root of unity, then there are three values for y; one is given by Eq. (3); the other two are ωy and $\omega^2 y$. The corresponding "mates" are

$$-q/3\omega y = (1/\omega)z = \omega^2 z$$

and

$$-q/3\omega^2 y = (1/\omega^2)z = \omega z.$$

We conclude that the roots of the cubic polynomial are given by the *cubic formula*:

$$y + z; \quad \omega y + \omega^2 z; \quad \omega^2 y + \omega z;$$

here

$$y^3 = \tfrac{1}{2}(-r + \sqrt{R})$$

and $R = r^2 + 4q^3/27$.

Example 16. If $f(x) = x^3 - 15x - 126$, then $f(x)$ is reduced [otherwise, one would reduce it via the substitution $x \mapsto x - b/3$]. Here, $q = -15$, $r = -126$, $R = 15376$, and $\sqrt{R} = 124$. Hence,

$$y^3 = \tfrac{1}{2}[-(-126) + 124] = 125,$$

so that $y = 5$; moreover, $z = -q/3y = 15/15 = 1$, so that one root is

$$u = y + z = 6.$$

The other two roots can be found either by using the quadratic formula on $(x^3 - 15x - 126)/(x - 6) = x^2 + 6x + 21$ (they are $-3 \pm 2i\sqrt{3}$) or by using the cubic formula (they now appear as $5\omega + \omega^2$ and $5\omega^2 + \omega$).

Example 17. Consider

$$f(x) = x^3 - 7x + 6 = (x - 1)(x - 2)(x + 3),$$

whose roots are, plainly, 1, 2, and -3. The cubic formula gives

$$y^3 = \tfrac{1}{2}\left(-6 + \sqrt{\tfrac{-400}{27}}\right),$$

and so one root of $f(x)$ is

$$\sqrt[3]{\tfrac{1}{2}\left(-6 + \sqrt{\tfrac{-400}{27}}\right)} + \sqrt[3]{\tfrac{1}{2}\left(-6 - \sqrt{\tfrac{-400}{27}}\right)}.$$

This expression is, thus, equal to 1, 2, or -3! It is not obvious that the value of the expression is real or rational, let alone an integer.

It is plain that a similar phenomenon will occur whenever $R = r^2 + 4q^3/27$ is negative. Every cubic has a real root, and the cubic formula involves $\sqrt{R}$.

Our 16th century ancestors were mystified by the phenomenon illustrated in Example 17. At that time, imaginary roots of quadratics (indeed, even negative roots of quadratics) were generally ignored. For example, to find the sides a and b of a rectangle having area A and perimeter p, one converts the equations

$$A = ab$$
$$p = 2a + 2b,$$

into the quadratic equation $2a^2 - pa + 2A = 0$ having roots

$$a = \tfrac{1}{4}\left(p \pm \sqrt{p^2 - 16A}\right).$$

If $p^2 - 16A$ is negative, then it is natural to say that there is no rectangle having the given area and perimeter. One would not invent complex numbers to find some ethereal rectangle living somewhere beyond the realm of the senses. But how can one explain square roots of negative numbers occurring in Example 17? The importance of the cubic formula in the history of mathematics is that such examples forced our ancestors to deal with complex numbers. We shall return to this point when we discuss the *Casus Irreducibilis* (Theorem 102).

Remark. There is a trigonometric solution to the cubic, due to Viète, that does give the roots of $f(x) = x^3 + qx + r$ in recognizable form (there is a proof in [Rotman, *A First Course in Abstract Algebra*]).
    If all the roots of $f(x)$ are real, then they are

$$t\cos(\alpha/3), \quad t\cos(\alpha/3 + 2\pi/3), \quad t\cos(\alpha/3 + 4\pi/3),$$

where $t = \sqrt{-4q/3}$ [q must be negative in this case] and $\cos(\alpha) = -4r/t^3$.
    If $f(x)$ has complex roots, then there are two possibilities, depending on the sign of $-4q/3$. If $-4q/3 \geq 0$, then the real root of $f(x)$ is

$$t\cosh(\beta/3),$$

where $\cosh(\beta) = -4r/t^3$. If $-4q/3 < 0$, then the real root of $f(x)$ is

$$t\sinh(\gamma/3),$$

where $\sinh(\gamma) = -4r/t^3$.

The **quartic formula** was found by Luigi Ferrari (ca. 1545), but we present the method of Descartes (1637). Consider the quartic polynomial

$$X^4 + aX^3 + bX^2 + cX + d;$$

setting $X = x - a/4$, we obtain a reduced polynomial

$$h(x) = x^4 + qx^2 + rx + s.$$

By Lemma 43, a formula for the roots of $h(x)$ will give a formula for the roots of the original one. Write

$$x^4 + qx^2 + rx + s = (x^2 + kx + \ell)(x^2 - kx + m),$$

where k, ℓ, and m are to be determined (the linear term in the second factor is $-k$ because $h(x)$ has no cubic term). If k, ℓ, and m are known, then the problem is solved by applying the quadratic formula. Expanding the right side and equating coefficients of like terms gives:

$$\ell + m - k^2 = q;$$
$$k(m - \ell) = r;$$
$$\ell m = s.$$

The first two equations yield:

$$2m = k^2 + q + r/k;$$
$$2\ell = k^2 + q - r/k.$$

Substituting these values of m and ℓ into the third equation gives

$$k^6 + 2qk^4 + (q^2 - 4s)k^2 - r^2 = 0.$$

This is a cubic in k^2 (essentially the "resolvent cubic" we will meet later), and one can thus solve for k^2 using the cubic formula. It is now easy to determine k, ℓ, and m, and hence to determine the roots of $h(x)$.

Example 18. It is not easy to produce an example of a quartic whose roots are given by the quartic formula in recognizable form. Here is one that I found in a 19th century textbook. If

$$f(x) = x^4 - 2x^2 + 8x - 3,$$

then the method leads to the cubic in k^2:

$$k^6 - 4k^4 + 16k^2 - 64 = k^6 - 2^2k^4 + 24k^2 - 2^6 = (k^2 - 2^2)^3,$$

so that $k = \pm 2$. It is now routine to find $\ell = -1$ and $m = 3$ if $k = 2$ (and $\ell = 3$ and $m = -1$ if $k = -2$); the reader can complete the calculation if desired. The roots are:

$$-1 + i\sqrt{2}, \quad -1 - i\sqrt{2}, \quad 1 + i\sqrt{2}, \quad 1 - i\sqrt{2}.$$

We can now see why our ancestors were tempted to find a similar formula for a quintic; surely it, too, would yield to ingenuity.

Exercises

68. Given numbers u and v, prove that there exist (possibly complex) numbers y and z such that

$$y + z = u \text{ and } yz = v.$$

69. Factor $x^3 + x^2 - 36$ in $\mathbb{Q}[x]$.

70. Let $g(x) = x^3 + qx + r$ and define $R = r^2 + 4q^3/27$. Let u be a root of $g(x)$ and let $u = y + z$, where $y^3 = \frac{1}{2}\left(-r + \sqrt{R}\right)$. Prove that

$$z^3 = \frac{1}{2}\left(-r - \sqrt{R}\right).$$

71. Find the roots of the following polynomials $f(x) \in \mathbb{R}[x]$.

(i) $f(x) = x^3 - 3x + 1$.

(ii) $f(x) = x^3 - 9x + 28$.

(iii) $f(x) = x^3 - 24x^2 - 24x - 25$.

(iv) $f(x) = x^3 - 15x - 4$.

(v) $f(x) = x^3 - 6x + 4$.

(vi) $f(x) = x^4 - 15x^2 - 20x - 6$.

Splitting Fields

Given a polynomial $f(x)$ with coefficients in a field F, we are going to describe the smallest field containing F and all the roots of $f(x)$.

Definition. If F is a subfield of a field E, one also says that E is a *field extension* of F, and one writes E/F is a field extension.

The notation E/F is designed to display the subfield F; it does not mean forming quotients (nor should it; after all, fields have no "honest" ideals).

Note that the term "extension" inverts one's viewpoint. Instead of focusing on subfields F of E, we focus on larger fields E containing F.

The following result will be useful.

Lemma 44. *Let E/F be a field extension, let $\alpha \in E$, and let $p(x) \in F[x]$ be a monic irreducible having α as a root.*

(i) $\partial(p) \le \partial(f)$ for every $f(x) \in F[x]$ having α as a root.

(ii) $p(x)$ is the only monic polynomial in $F[x]$ of degree $\partial(p)$ that has α as a root.

Proof. (i) Let
$$I = \{f(x) \in F[x] : f(\alpha) = 0\}.$$
It is easy to see that I is an ideal: if $f(x), g(x) \in I$, then $f(\alpha) + g(\alpha) = 0 + 0 = 0$ and $f + g \in I$; if $f(x) \in I$ and $h(x) \in F[x]$, then $h(\alpha)f(\alpha) = h(\alpha) \cdot 0 = 0$ and $hf \in I$.

If $f(x) \in I$, then $d = (f, p) \in I$, for d is a linear combination of f and p. Since p is irreducible, its only monic divisors are 1 and p, and so $d = 1$ or $d = p$. But $1 \notin I$, because α is not a root of the constant polynomial 1, so that $d = p$ and $p \mid f$. Therefore, $\partial(p) \le \partial(f)$.

(ii) If $q(x)$ is a monic polynomial in I with $\partial(q) = \partial(p)$, then $q - p$ is a polynomial in I. If $q - p \ne 0$, then it has a degree and $\partial(q - p) < \partial(p)$, contradicting the inequality in part (i).   •

We have already observed that if F is a subfield of E, then E may be viewed as a vector space over F: scalar multiplication is defined by letting $c\alpha$, for $c \in F$ and $\alpha \in E$, be the product of the two elements c and α under the given multiplication on E.

Definition. The dimension of E viewed as a vector space over F is called the *degree* of E over F and it is denoted by $[E : F]$. One says that E/F is a *finite extension* if $[E : F]$ is finite.

The next theorem explains why $[E : F]$ is called the degree.

Theorem 45. *Let $p(x) \in F[x]$ be an irreducible polynomial of degree d. Then $E = F[x]/(p(x))$ is a field extension of F of degree d.*

Indeed, E contains a root α of $p(x)$, and a basis of E as a vector space over F is $\{1, \alpha, \alpha^2, \ldots, \alpha^{d-1}\}$.

Proof. Denote $(p(x))$ by I, and denote $x + I$ in E by α; it suffices to prove that $\{1, \alpha, \alpha^2, \ldots, \alpha^{d-1}\}$ is a basis of E over F (we continue to identify each $c \in F$ with $c + I \in F[x]/I$). If, for $0 \leq i \leq d - 1$, there are $c_i \in F$, not all 0, with $\sum c_i \alpha^i = 0$, then α is a root of $f(x) = \sum c_i x^i$, a polynomial of degree $< d$, contradicting $p(x)$ being a polynomial in $F[x]$ of least degree having α as a root (Lemma 44). Hence $\{1, \alpha, \alpha^2, \ldots, \alpha^{d-1}\}$ is linearly independent. To see that this set spans E, note first that every element of E has the form $f(x) + I$. The division algorithm gives $q(x)$ and $r(x)$ with $f(x) = q(x)p(x) + r(x)$, where $\partial(r) < \partial(p) = d$, and so $f(x) + I = r(x) + I$. Hence $\{1, \alpha, \alpha^2, \ldots, \alpha^{d-1}\}$ is a basis of E.   •

Definition. Let E/F be a field extension, and let $\alpha_1, \ldots, \alpha_n \in E$. Then $F(\alpha_1, \ldots, \alpha_n)$, called the field obtained by *adjoining* $\alpha_1, \ldots, \alpha_n$ to F, is the intersection of all the subfields of E which contain F and $\{\alpha_1, \ldots, \alpha_n\}$. An extension E/F is a *simple extension* if it is obtained by adjoining just one element α to F; that is,

$$E = F(\alpha) = \{f(\alpha)/g(\alpha) : f(x), g(x) \in F[x] \text{ and } g(\alpha) \neq 0\}.$$

It is easy to see that $F(\alpha_1, \ldots, \alpha_n)$ is the smallest subfield of E containing F and $\{\alpha_1, \ldots, \alpha_n\}$ in the sense that $F(\alpha_1, \ldots, \alpha_n) \subset K$ for any other such subfield K.

Definition. Let E/F be a field extension, and let $\alpha \in E$. Then α is *algebraic over F* if α is a root of some monic polynomial $\in F[x]$; otherwise α is *transcendental over F*. A field extension E/F is called *algebraic* if every element of E is algebraic over F.

When one says that π or e is transcendental, one usually means that they are transcendental over $\mathbb{Q}$. Recall that when F is a field, $E = F(x) = \text{Frac}(F[x])$ denotes the field of all rational functions over F; its elements are all $f(x)/g(x)$, where $f(x), g(x) \in F[x]$ and $g(x) \neq 0$. In this case, $x \in F(x)$ is transcendental over F.

Theorem 46. *If E/F is a finite extension, then it is an algebraic extension.*

Proof. Assume that $[E : F] = n$ and that $\alpha \in E$. In any n-dimensional vector space, any sequence of $n + 1$ vectors is linearly dependent. There are thus scalars $c_i \in F$ for $i = 0, 1, \ldots, n$, not all 0, with

$$\sum_{i=0}^{n} c_i \alpha^i = 0;$$

there is thus a nonzero polynomial in $F[x]$ having α as a root, and so α is algebraic over F. •

The converse of this theorem is false. In Exercise 72, we give an algebraic extension $E/\mathbb{Q}$ that is not finite.

Theorem 47. *Let E/F be a field extension, and let $\alpha \in E$ be algebraic over F.*

(i) *There is a monic irreducible polynomial $p(x) \in F[x]$ having α as a root;*

(ii) *$F[x]/(p(x)) \cong F(\alpha)$; in fact, there is an isomorphism*

$$\Phi : F[x]/(p) \to F(\alpha),$$

fixing F pointwise, with $\Phi(x + (p)) = \alpha$.

(iii) *$p(x)$ is the unique monic polynomial of least degree in $F[x]$ having α as a root;*

(iv) *$[F(\alpha) : F] = \partial(p)$.*

Proof. (i) Define $\varphi : F[x] \to E$ to be the function $f(x) \mapsto f(\alpha)$; it is a ring map because it is the restriction of the evaluation map $e_\alpha : E[x] \to E$ to $F[x]$. Now $\ker \varphi$ is a nonzero ideal in $F[x]$ because α is algebraic over F; as $F[x]$ is a PID, $\ker \varphi = (p(x))$ for some monic polynomial $p(x) \in F[x]$. Since E is a field, $\operatorname{im} \varphi$ is a domain. By the first isomorphism theorem, $F[x]/\ker \varphi \cong \operatorname{im} \varphi$, so that $\ker \varphi = (p(x))$ is a prime ideal. Therefore, $p(x)$ is an irreducible polynomial in $F[x]$, by Theorem 24.

(ii) The first isomorphism theorem says the map $\Phi : F[x]/(p) \to \operatorname{im} \varphi$, given by $f(x) + (p) \mapsto f(\alpha)$, is an isomorphism; thus $\Phi : x + (p) \mapsto \alpha$ and $\Phi : c + (p) \mapsto c$ for each $c \in F$. As always, we identify the subfield $F' = \{c+(p) : c \in F\}$ with F, and so we may say that Φ fixes F pointwise. Finally,

$$\operatorname{im} \varphi = \operatorname{im} \Phi = \{f(\alpha) : f(x) \in F[x]\}$$

is a subfield of E, by Corollary 29, because $p(x)$ is irreducible. Clearly, im Φ is contained in any subfield of E which contains F and α, so that im $\Phi = F(\alpha)$.

(iii) This is Lemma 44.

(iv) This is Theorem 45. •

Definition. The polynomial $p(x)$ in Theorem 47 is called the ***irreducible polynomial of α over F***.

When α is algebraic over F, the general description of $F(\alpha)$ as being comprised of rational functions in α simplifies to $F(\alpha)$ being comprised of polynomials in α. In particular, we have seen in Example 11 that the multiplicative inverse of $f(\alpha)$ is $s(\alpha)$, where $s(x)f(x) + t(x)p(x) = 1$ and $p(x)$ is the irreducible polynomial of α.

Definition. A ***splitting field*** of $f(x) \in F[x]$ is a field extension E/F in which $f(x)$ splits (it is a product of linear factors) while $f(x)$ does not split in any proper subfield of E.

Example 19. If $\omega \neq 1$ is a cube root of unity, then $x^3 - 1 \in \mathbb{Q}[x]$ splits over $\mathbb{C}$, but its splitting field is $\mathbb{Q}(\omega)$.

Theorem 48. *If F is a field, then every polynomial $f(x) \in F[x]$ has a splitting field.*

Proof. By Kronecker's theorem (Theorem 30), there is a field extension K/F over which $f(x)$ splits. Let $\alpha_1, \ldots, \alpha_n$ be the roots of $f(x)$ in K, and define $E = F(\alpha_1, \ldots, \alpha_n)$. It is plain that $f(x)$ splits over E, and $f(x)$ does not split over any proper subfield of E (which necessarily omits one of the α_i). •

Notice that the splitting field just constructed depends on a choice of field K in Kronecker's theorem, and so it is not unique. For example, here are two isomorphic copies of $\mathbb{C}$: all ordered pairs of real numbers $(a, b) = a + bi$; all cosets in $\mathbb{R}[x]/(x^2 + 1)$; each of these is a splitting field of $f(x) = x^2 + 1$ over $\mathbb{R}$. We will soon prove that any two splitting fields of a polynomial over a field F are isomorphic.

The following degree formula is very useful.

Lemma 49   (Degree Formula.). *If $F \subset B \subset E$ are fields with $[E : B]$ and $[B : F]$ finite, then E/F is finite and*

$$[E : F] = [E : B][B : F].$$

Proof. Let $\{\alpha_1, \ldots, \alpha_m\}$ be a basis of E/B, and let $\{\beta_1, \ldots, \beta_n\}$ be a basis of B/F. It suffices to prove that $\{\beta_j \alpha_i : 1 \leq i \leq m, 1 \leq j \leq n\}$ is a basis of E/F.

This set spans E. If $\gamma \in E$, then there are b_i in B with $\gamma = \sum b_i \alpha_i$. But each $b_i = \sum c_{ij} \beta_j$ for c_{ij} in F; hence $\gamma = \sum c_{ij} \beta_j \alpha_i$. To see that this set is linearly independent, assume that $\sum c_{ij} \beta_j \alpha_i = 0$ for c_{ij} in F. Now $b_i = \sum c_{ij} \beta_j \in B$, so that independence of the α_i over B implies that $b_i = 0$ for all i. Hence $\sum c_{ij} \beta_j = 0$ for all i, and so the independence of the β_j over F implies that $c_{ij} = 0$ for all i, j, as desired. •

Example 20. Let $E = \mathbb{Q}(\sqrt{2}, \sqrt{3})$ and let $B = \mathbb{Q}(\sqrt{2})$. Now $\sqrt{3}$ is algebraic over $\mathbb{Q}$ and its irreducible polynomial is $x^2 - 3$; it follows that $\sqrt{3}$ is algebraic over B. Moreover, the irreducible polynomial $p(x)$ of $\sqrt{3}$ over $\mathbb{Q}(\sqrt{2})$ is a divisor of $x^2 - 3$, so that $[E : \mathbb{Q}(\sqrt{2})] \leq 2$. Indeed, $[E : \mathbb{Q}(\sqrt{2})] = 2$ because $\sqrt{3} \notin \mathbb{Q}(\sqrt{2})$ (otherwise, there are rationals a and b with $\sqrt{3} = a + b\sqrt{2}$, so that $3 = a^2 + 2ab\sqrt{2} + 2b^2$, which contradicts the irrationality of $\sqrt{2}$). Therefore, Lemma 49 gives

$$[E : \mathbb{Q}] = [E : \mathbb{Q}(\sqrt{2})][\mathbb{Q}(\sqrt{2}) : \mathbb{Q}] = 4.$$

Define $\alpha = \sqrt{2} + \sqrt{3} \in E$, and note that α is algebraic over $\mathbb{Q}$ because $E/\mathbb{Q}$ is an algebraic extension, by Theorem 46. What is the irreducible polynomial of α?

$$\alpha^2 = \left(\sqrt{2} + \sqrt{3}\right)^2 = 5 + 2\sqrt{6},$$

so that

$$\alpha^2 - 5 = 2\sqrt{6},$$

and hence

$$\alpha^4 - 10\alpha^2 + 1 = 0.$$

By Exercise 67, the polynomial $p(x) = x^4 - 10x^2 + 1$ is irreducible over $\mathbb{Q}$. We have seen above that $[E : \mathbb{Q}] = 4$. On the other hand, $\mathbb{Q}(\alpha) \subset E$, so that Theorem 47 gives $[\mathbb{Q}(\alpha) : \mathbb{Q}] = 4$; by Lemma 49,

$$4 = [E : \mathbb{Q}] = [E : \mathbb{Q}(\alpha)][\mathbb{Q}(\alpha) : \mathbb{Q}] = 4[E : \mathbb{Q}(\alpha)];$$

hence, $[E : \mathbb{Q}(\alpha)] = 1$ and so $E = \mathbb{Q}(\alpha)$. We have shown that $E/\mathbb{Q}$ is a simple extension.

What is α^{-1}? Since $\alpha^4 - 10\alpha^2 + 1 = 0$, we have $\alpha(10\alpha - \alpha^3) = 1$ so that $\alpha^{-1} = 10\alpha - \alpha^3$. Replacing α by $\sqrt{2} + \sqrt{3}$, one can write α^{-1} explicitly in terms of $\sqrt{2}$ and $\sqrt{3}$.

Lemma 50. *Let $\sigma : F \to F'$ be an isomorphism of fields, let $\sigma^* : F[x] \to F'[x]$, defined by $\sum r_i x^i \mapsto \sum \sigma(r_i)x^i$, be the corresponding isomorphism of rings, let $p(x) \in F[x]$ be irreducible, and let $p^*(x) = \sigma^*(p(x)) \in F'[x]$.*

If β is a root of $p(x)$ and β' is a root of $p^*(x)$, then there is a unique isomorphism $\hat{\sigma} : F(\beta) \to F'(\beta')$ extending σ with $\hat{\sigma}(\beta) = \beta'$.

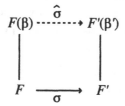

Proof. The isomorphism $\sigma^* : F[x] \to F'[x]$ carries the ideal $(p(x))$ onto the ideal $(p^*(x))$, and so Exercise 39 provides an isomorphism $\Sigma : F[x]/(p) \to F'[x]/(p^*)$ with $c + (p) \mapsto \sigma(c) + (p^*)$ for all $c \in F$ and $x + (p) \mapsto x + (p^*)$. Define $\hat{\sigma}$ as the composite

$$F(\beta) \xrightarrow{\Phi^{-1}} F[x]/(p) \xrightarrow{\Sigma} F'[x]/(p^*) \xrightarrow{\Phi'} F'(\beta').$$

Using Theorem 47, it is easy to see that $\hat{\sigma}$ is an isomorphism extending σ that sends β to β'. The uniqueness of $\hat{\sigma}$ follows from Exercise 73. ●

We now extend Lemma 50 so that it treats not necessarily irreducible polynomials. The second part of it introduces a new kind of polynomial.

Definition. Let $f(x) \in F[x]$ have the factorization into (not necessarily distinct) irreducibles:

$$f(x) = ap_1(x) \cdots p_t(x),$$

where $a \in F$; then $f(x)$ is **separable** if each $p_i(x)$ has no repeated roots.

Let F be a field and let $q(x) \in F[x]$ be irreducible. If the derivative $q'(x)$ is not the zero polynomial, then its degree is smaller than the degree of $q(x)$; hence $(q, q') = 1$ and $q(x)$ is separable, by Exercise 44. It follows that if F has characteristic 0, then every nonconstant polynomial is separable; if F has characteristic p, then it is possible that $q' = 0$ (see Example 21 below). Fields in which every nonconstant polynomial is separable are called **perfect**. Thus, every field of characteristic 0 is perfect; Exercise 77 asks the reader to prove that every finite field is perfect.

Definition. If E/F is an extension, then $\alpha \in E$ is called *separable* if either it is transcendental or its irreducible polynomial is separable; an extension is called *separable* if every one of its elements is separable.

Example 21. Here is an example of an inseparable extension. Let $K = \mathbb{Z}_p(t)$, the field of all rational functions over $\mathbb{Z}_p$. The polynomial $q(x) = x^p - t \in K[x]$ is irreducible over K (see Exercise 75). Its splitting field E/K is not separable: if $\alpha \in E$ is a root of $q(x)$, then $x^p - t = (x - \alpha)^p$ in $E[x]$ because E has characteristic p. Note that $q'(x) = px^{p-1} = 0$.

Remark. When studying infinite fields of characteristic $p > 0$, it is important to know whether a field extension E/F is separable. One can prove that

$$E_s = \{a \in E : a \text{ is separable over } F\},$$

called the *separable closure* of F in E, is a subfield of E. An element $a \in E$ is called *purely inseparable* if its irreducible polynomial factors as $(x - \alpha)^m$ over some splitting field (m must be a power of p); an extension K/F is called *purely inseparable* if every $a \in K$ is purely inseparable over F. Now E/E_s is purely inseparable, and so every extension E/F is a separable extension E_s/F followed by a purely inseparable extension E/E_s. When E/F is finite, then the degrees of these extensions are useful (for proofs of these results, see van der Waerden, *Modern Algebra*).

Theorem 51. *Let $\sigma : F \to F'$ be an isomorphism of fields, let $f(x) \in F[x]$, and let $f^*(x) = \sigma^*(f(x))$ be the corresponding polynomial in $F'[x]$; let E be a splitting field of $f(x)$ over F and let E' be a splitting field of $f^*(x)$ over F'.*

   (i) *There is an isomorphism $\tilde{\sigma} : E \to E'$ extending σ.*

   (ii) *If $f(x)$ is separable, then σ has exactly $[E : F]$ extensions $\tilde{\sigma}$.*

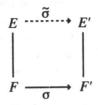

Proof. (i) The proof is by induction on $[E : F]$. If $[E : F] = 1$, then $E = F$ and $f(x)$ is a product of linear factors in $F[x]$; it follows that $f^*(x)$ is also a product of linear factors, and so $E' = F'$; therefore, we may define $\tilde{\sigma} = \sigma$. If $[E : F] > 1$, choose an irreducible factor $p(x)$ of $f(x)$ having degree ≥ 2, and choose a root β of $p(x)$, hence a root of $f(x)$, which must be in E. Let $p^*(x) \in F'[x]$ correspond to $p(x)$, and let $\beta' \in E'$ be a root of $p^*(x)$. By Lemma 50, for each such β' there is a unique isomorphism $\hat{\sigma} : F(\beta) \to F'(\beta')$ extending σ with $\hat{\sigma}(\beta) = \beta'$. Now E is a splitting field of $f(x)$ over $F(\beta)$ and E' is a splitting field of $f^*(x)$ over $F'(\beta')$. Since $[E : F] = [E : F(\beta)][F(\beta) : F]$, and since $[F(\beta) : F] \geq 2$, it follows that $[E : F(\beta)] < [E : F]$. By induction, there exists $\tilde{\sigma} : E \to E'$ extending $\hat{\sigma}$, hence extending σ.

(ii) This proof, a modification of that in part (i), also proceeds by induction on $[E : F]$. If $[E : F] = 1$, then $E = F$ and there is only one extension $\tilde{\sigma}$ of σ, namely, σ itself. If $[E : F] > 1$, let $f(x) = p(x)g(x)$, where $p(x)$ is irreducible of degree d, say. If $d = 1$, then we may replace $f(x)$ by $g(x)$ without changing the problem. If $d > 1$, choose a root β of $p(x)$. If $\tilde{\sigma}$ is any extension of σ to E, then $\tilde{\sigma}(\beta)$ is a root β' of $p^*(x)$; since $f^*(x)$ is separable, $p^*(x)$ has exactly d roots $\beta' \in E'$; by Lemma 50, there are exactly d isomorphisms $\hat{\sigma} : F(\beta) \to F'(\beta')$ extending σ, one for each β'. Now E is a splitting field of $f(x)$ over $F(\beta)$ and E' is a splitting field of $f^*(x)$ over $F'(\beta')$. Since $[E : F(\beta)] = [E : F]/d$, induction shows that each of the d isomorphisms $\hat{\sigma}$ has exactly $[E : F]/d$ extensions to E; therefore, σ has exactly $[E : F]$ extensions $\tilde{\sigma}$, because every τ extending σ has $\tau|F(\beta) = $ some $\hat{\sigma}$.   •

Corollary 52. *If $f(x) \in F[x]$, then any two splitting fields of $f(x)$ over F are isomorphic by an isomorphism fixing F pointwise.*

Proof. In Theorem 51(i), choose $F = F'$ and σ the identity on F.   •

Corollary 53 (E.H. Moore). *Any two finite fields of order $q = p^n$ are isomorphic.*

Proof. Any field F of order q is the splitting field of $x^q - x$ over $\mathbb{Z}_p$, as we saw in Theorem 33.   •

One calls *the* field of order p^n the *Galois field* of this order and denotes it by $GF(p^n)$, although $GF(p)$ is usually denoted by $\mathbb{Z}_p$. Another common notation for the field with $q = p^n$ elements is $\mathbb{F}_q$.

Both $x^3 + x + 1$ and $x^3 + x^2 + 1$ are irreducible in $\mathbb{Z}_2[x]$, for neither has a root in $\mathbb{Z}_2$. Hence, $\mathbb{Z}_2[x]/(x^3 + x + 1)$ and $\mathbb{Z}_2[x]/(x^3 + x^2 + 1)$ are fields of degree 3 over $\mathbb{Z}_2$; that is, both fields have $2^3 = 8$ elements. By Moore's theorem, both of these fields are isomorphic. More generally, one sees that if $f(x)$ and $g(x)$ are irreducible polynomials over $\mathbb{Z}_p$ which have the same degree, then $\mathbb{Z}_p[x]/(f(x))$ and $\mathbb{Z}_p[x]/(g(x))$ are isomorphic.

Exercises

72. (i) Let E/F be an extension, and let $\alpha, \beta \in E$ be algebraic elements over F. If $\alpha \neq 0$, prove that $\alpha + \beta$, $\alpha\beta$, and α^{-1} are all algebraic over F. (Hint. Use Lemma 49 to prove that $F(\alpha, \beta)$ is a finite-dimensional vector space over F.)

  (ii) If E/F is an extension, prove that the subset

$$K = \{\alpha \in E : \alpha \text{ is algebraic over } F\}$$

is a subfield of E containing F.

  (iii) Define the **algebraic numbers** $\mathbb{A}$ to be the set of all those complex numbers that are algebraic over $\mathbb{Q}$. Prove that $\mathbb{A}/\mathbb{Q}$ is an algebraic extension that is not finite.

73. Let F be a field. Prove that if σ is an isomorphism of $F(\alpha_1, \ldots, \alpha_n)$ with itself such that $\sigma(\alpha_i) = \alpha_i$ for $i = 1, \ldots, n$, and $\sigma(c) = c$ for all $c \in F$, then σ is the identity. Conclude that if E is a field extension of F and if $\sigma, \tau : F(\alpha_1, \ldots, \alpha_n) \to E$ fix F pointwise and $\sigma(\alpha_i) = \tau(\alpha_i)$ for all i, then $\sigma = \tau$.

74. If $F \subset B \subset E$ are fields and E/F is finite, then both E/B and B/F are finite, and $[E : F] = [E : B][B : F]$.

75. If $K = \mathbb{Z}_p(t)$, prove that $f(x) = x^p - t$ is irreducible in $K[x]$. (Hint. If E/K is a splitting field of $f(x)$, then $x^p - t = (x - \alpha)^p$ for some $\alpha \in E$.)

76. Show that a field F of characteristic p is perfect if and only if every element of F has a pth root in F.

77. Show that every finite field F is perfect. (Hint: The function $a \mapsto a^p$ is always an injection $F \to F$.)

The Galois Group

We now set up an analogy with symmetries of polygons in the plane even though some of the algebraic analogues have not yet been defined.

polygon P	polynomial $f(x) \in F[x]$
plane[7]	splitting field E of $f(x)$
$\text{Vert}(P) = \{v_1, \ldots, v_n\}$	roots $\alpha_1, \ldots, \alpha_n$
linear transformation	automorphism of E
orthogonal transformation	automorphism of E fixing F
$\Sigma(P)$	Galois group $\text{Gal}(f) = \text{Gal}(E/F)$
regular polygon[8]	irreducible polynomial

In the geometric setting, we saw that if P has n vertices, then $\Sigma(P)$ is isomorphic to a subgroup of S_n, but we did not, in general, compute $|\Sigma(P)|$ more precisely. We did see, in Theorem 4, that different types of triangles have nonisomorphic symmetry groups.

Definition. If E is a field, then an *automorphism* of E is an isomorphism of E with itself. If E/F is a field extension, then an automorphism σ of E *fixes F pointwise* if $\sigma(c) = c$ for every $c \in F$.

The next lemma, though very easy to prove, is fundamental; it is the analogue of Lemma 2.

Lemma 54. *Let $f(x) \in F[x]$ and let E/F be an extension field of F. If $\sigma : E \to E$ is an automorphism fixing F pointwise, and if $\alpha \in E$ is a root of $f(x)$, then $\sigma(\alpha)$ is also a root of $f(x)$.*

Proof. Let $f(x) = c_0 + c_1 x + \cdots + c_n x^n$, so that

$$c_0 + c_1 \alpha + \cdots + c_n \alpha^n = 0.$$

Applying σ gives

$$\sigma(c_0) + \sigma(c_1)\sigma(\alpha) + \cdots + \sigma(c_n)\sigma(\alpha)^n$$
$$= c_0 + c_1\sigma(\alpha) + \cdots + c_n\sigma(\alpha)^n = 0,$$

because σ fixes F. Therefore, $\sigma(\alpha)$ is a root of $f(x)$.   •

---

[7] Since splitting fields of various polynomials do not all have the same dimension, this analogy can be improved by considering polyhedra in higher dimensional euclidean space as well as polygons in the plane.

[8] See Exercise 2 and Exercise 79.

Definition.[9] Let E/F be a field extension. Its *Galois group* is

$$\text{Gal}(E/F) = \{\text{automorphisms } \sigma \text{ of } E \text{ fixing } F \text{ pointwise}\}$$

under the binary operation of composition. If $f(x) \in F[x]$ has splitting field E, then the *Galois group* of $f(x)$ is $\text{Gal}(E/F)$.

It is easy to check that $\text{Gal}(E/F)$ is a group; it is a subgroup of the group of all automorphisms of E.

Theorem 55. *If $f(x) \in F[x]$ has n distinct roots in its splitting field E, then $\text{Gal}(E/F)$ is isomorphic to a subgroup of the symmetric group S_n, and so its order is a divisor of $n!$.*

Proof. Let $X = \{\alpha_1, \dots, \alpha_n\}$ be the set of all the roots of $f(x)$ in E. By Lemma 54, if $\sigma \in \text{Gal}(E/F)$, then $\sigma(X) = X$. The map $\text{Gal}(E/F) \to S_X$ defined by $\sigma \mapsto \sigma|X$ is easily seen to be a homomorphism; it is an injection, by Exercise 73. Finally, $S_X \cong S_n$. •

For example, the Galois group of a quartic polynomial is a subgroup of S_4, and the Galois group of a quintic polynomial is a subgroup of S_5.

Example 22. The splitting field of $x^2 + 1$ over $\mathbb{R}$ is, of course, $\mathbb{C}$, and $|\text{Gal}(\mathbb{C}/\mathbb{R})| \leq 2$, by Theorem 55. In fact, $|\text{Gal}(\mathbb{C}/\mathbb{R})| = 2$ because the group contains the automorphism

$$\sigma : z = a + ib \mapsto \bar{z} = a - ib.$$

Notice that $\sigma : i \mapsto -i$ and $-i \mapsto i$, and so it interchanges the roots. One sees that the elements of the Galois group should be regarded as generalizations of complex conjugation.

Theorem 56. *If $f(x) \in F[x]$ is a separable polynomial and if E/F is its splitting field, then*
$$|\text{Gal}(E/F)| = [E : F].$$

Proof. By Theorem 51(ii) with $F = F'$, $E = E'$, and $\sigma : F \to F$ the identity, there are exactly $[E : F]$ automorphisms of E that fix F. •

Since $x^2 + 1$ is separable and $[\mathbb{C} : \mathbb{R}] = 2$, we see once again that $|\text{Gal}(\mathbb{C}/\mathbb{R})| = 2$.

---

[9]This is not the definition of Galois; it is the modern version introduced by E. Artin around 1930, and it is isomorphic to the original version.

Example 23. Let $f(x) = x^3 - 1 \in \mathbb{Q}[x]$; $f(x)$ is separable, because $\mathbb{Q}$ has characteristic 0. Now $f(x) = (x - 1)(x^2 + x + 1)$ is a factorization of $f(x)$ into irreducibles. If E is the splitting field of $f(x)$ over $\mathbb{Q}$, then $E = \mathbb{Q}(1, \omega, \omega^2) = \mathbb{Q}(\omega)$, where ω is a primitive cube root of unity, that is, ω is a root of $x^2 + x + 1$. Since $x^2 + x + 1$ is an irreducible polynomial of degree 2, we have

$$2 = [\mathbb{Q}(\omega) : \mathbb{Q}] = [E : \mathbb{Q}] = |\operatorname{Gal}(E/\mathbb{Q})|,$$

by Theorem 56. Therefore, the Galois group is cyclic of order 2. Its generator takes $\omega \mapsto \omega^2 = \overline{\omega}$; that is, the generator is complex conjugation.

Example 24. Let $g(x) = x^3 - 2 \in \mathbb{Q}[x]$. The roots of $g(x)$ are $\alpha, \omega\alpha$, and $\omega^2\alpha$, where $\alpha = \sqrt[3]{2}$ is the real cube root of 2 and $\omega = e^{2\pi i/3}$ is a primitive cube root of unity, and so the splitting field E of $g(x)$ is $E = \mathbb{Q}(\alpha, \omega\alpha, \omega^2\alpha)$. We claim that $E = \mathbb{Q}(\alpha, \omega)$: $E \subset \mathbb{Q}(\alpha, \omega)$ because α, $\omega\alpha, \omega^2\alpha \in \mathbb{Q}(\alpha, \omega)$; $\mathbb{Q}(\alpha, \omega) \subset E$ because $\omega = \omega\alpha/\alpha \in E$. Since $g(x)$ is irreducible over $\mathbb{Q}$, we have $[\mathbb{Q}(\alpha) : \mathbb{Q}] = 3$. But $\mathbb{Q}(\alpha)$ consists wholly of real numbers, and so it cannot be the splitting field E of $g(x)$. Hence, $[E : \mathbb{Q}(\alpha)] > 1$, and

$$|\operatorname{Gal}(E/\mathbb{Q})| = [E : \mathbb{Q}] = [E : \mathbb{Q}(\alpha)][\mathbb{Q}(\alpha) : \mathbb{Q}] = 3[E : \mathbb{Q}(\alpha)] > 3;$$

it follows that $\operatorname{Gal}(E/\mathbb{Q}) \cong S_3$, by Theorem 55.

Lemma 57. *Let $F \subset B \subset E$ be a tower of fields with B/F the splitting field of some polynomial $f(x) \in F[x]$. If $\sigma \in \operatorname{Gal}(E/F)$, then $\sigma|B \in \operatorname{Gal}(B/F)$.*

Proof. It suffices to prove that $\sigma(B) = B$. If $\alpha_1, \ldots, \alpha_n$ are the distinct roots of $f(x)$, then $B = F(\alpha_1, \ldots, \alpha_n)$. Now $\sigma(F) = F$, and $\sigma(\alpha_i) \in B$ for all i, by Lemma 54; it follows that

$$\sigma(B) = \sigma(F(\alpha_1, \ldots, \alpha_n)) = F(\sigma(\alpha_1), \ldots, \sigma(\alpha_n)) = B,$$

as desired.   ●

Theorem 58. *Let $F \subset B \subset E$ be a tower of fields with B/F the splitting field of some polynomial $f(x) \in F[x]$ and E/F the splitting field of some $g(x) \in F[x]$. Then $\operatorname{Gal}(E/B)$ is a normal subgroup of $\operatorname{Gal}(E/F)$, and*

$$\operatorname{Gal}(E/F)/\operatorname{Gal}(E/B) \cong \operatorname{Gal}(B/F).$$

Proof. Define $\psi : \mathrm{Gal}(E/F) \to \mathrm{Gal}(B/F)$ by $\sigma \mapsto \sigma|B$; Lemma 57 says that ψ does take its values in $\mathrm{Gal}(B/F)$. It is easily seen that ψ is a homomorphism with $\ker \psi = \mathrm{Gal}(E/B)$ [if $\sigma|B =$ identity, then σ is an automorphism of E fixing B], so that the latter is a normal subgroup of $\mathrm{Gal}(E/F)$. If $\tau \in \mathrm{Gal}(B/F)$, then Theorem 51 shows that there is an automorphism $\tilde{\tau}$ of E with $\psi(\tilde{\tau}) = \tilde{\tau}|B = \tau$. Hence ψ is surjective, and the first isomorphism theorem for groups gives the result.   ●

Remark. The hypothesis that E/F is a splitting field enters only in showing that ψ is surjective. Without this hypothesis, one can prove only that the quotient group is isomorphic to a subgroup of $\mathrm{Gal}(B/F)$.

Example 25. The Galois group of $f(x) = x^3 - 2 \in \mathbb{Q}[x]$ was computed in Example 24. If $\alpha = \sqrt[3]{2}$ and $\omega = e^{2\pi i/3}$, then we have seen that $E = \mathbb{Q}(\alpha, \omega)$ is a splitting field of $f(x)$ over $\mathbb{Q}$, that $[E : \mathbb{Q}] = 6$, and that $\mathrm{Gal}(E/\mathbb{Q}) \cong S_3$.

Let us now view this example in light of Theorem 58. Consider the tower of fields

$$\mathbb{Q} \subset \mathbb{Q}(\omega) \subset E.$$

Since $\mathbb{Q}(\omega)/\mathbb{Q}$ is a splitting field (of $x^2 + x + 1$), Theorem 58 gives

$$\mathrm{Gal}(E/\mathbb{Q}(\omega)) \lhd \mathrm{Gal}(E/\mathbb{Q})$$

and

$$\mathrm{Gal}(E/\mathbb{Q})/\mathrm{Gal}(E/\mathbb{Q}(\omega)) \cong \mathrm{Gal}(\mathbb{Q}(\omega))/\mathbb{Q}).$$

Now $\mathrm{Gal}(\mathbb{Q}(\omega)/\mathbb{Q})$ has order 2, because x^2+x+1 is an irreducible polynomial of degree 2 (it has no rational roots). We claim that $\mathrm{Gal}(E/\mathbb{Q}(\omega))$ has order 3. A moment's thought shows that none of the roots α, $\omega\alpha$, and $\omega^2\alpha$ of $x^3 - 2$ lies in $\mathbb{Q}(\omega)$. Since a cubic is irreducible over a field F if it has no roots in F, we see that $x^3 - 2$ is irreducible over $\mathbb{Q}(\omega)$. By Theorem 45, $[E : \mathbb{Q}(\omega)] = 3$, and by Theorem 56, $|\mathrm{Gal}(E/\mathbb{Q}(\omega))| = 3$. Therefore, in this case, the isomorphism of Theorem 58 is just $S_3/A_3 \cong \mathbb{Z}_2$. Note that $A_3 = \langle\sigma\rangle$, where $\sigma(\omega) = \omega$ and $\sigma : \alpha \mapsto \omega\alpha$. Hence, $\sigma(\omega\alpha) = \omega^2\alpha$ and $\sigma(\omega^2\alpha) = \alpha$, so that σ is a 3-cycle.

Exercises

78. Let $f(x) \in F[x]$ be an irreducible polynomial of degree n, and let E/F be a splitting field of $f(x)$.

    (i) Prove that $n \mid [E : F]$.

    (ii) Prove that if $f(x)$ is separable, then $n \mid |\mathrm{Gal}(E/F)|$.

79. Let $f(x) \in F[x]$, let E/F be a splitting field, and let $G = \mathrm{Gal}(E/F)$ be the Galois group.

    (i) If $f(x)$ is irreducible, then G acts *transitively* on the set of all roots of $f(x)$ (if α and β are any two roots of $f(x)$ in E, there exists $\sigma \in G$ with $\sigma(\alpha) = \beta$). (Hint: Lemma 50.)

    (ii) If $f(x)$ has no repeated roots and G acts transitively on the roots, then $f(x)$ is irreducible. Conclude, after comparing with Exercise 2, that irreducible polynomials are analogous to regular polygons. (Hint: If $f(x) = g(x)h(x)$, then the gcd $(g(x), h(x)) = 1$; if α is a root of $g(x)$ such that $\sigma(\alpha)$ is a root of $h(x)$, then $\sigma(\alpha)$ is a common root of $g(x)$ and $h(x)$.)

80. Let E be the splitting field of $f(x) = x^4 - 10x^2 + 1$ over $\mathbb{Q}$. Find $\mathrm{Gal}(E/\mathbb{Q})$. (Hint. See Exercise 67 and Example 20. The roots of $f(x)$ are

$$\sqrt{2} + \sqrt{3}, \quad \sqrt{2} - \sqrt{3}, \quad -\sqrt{2} + \sqrt{3}, \quad -\sqrt{2} - \sqrt{3}.)$$

Roots of Unity

The simplest field extensions of a field F are those in which we adjoin an nth root of an element $c \in F$. To investigate these, it will be valuable for us to consider roots of unity. After all, if $\alpha^n = c$, then the other nth roots of c are of the form $\omega\alpha$, where ω is some nth root of unity, and so it will be relevant whether or not F contains such roots of unity. Note that $\mathbb{R}$, for example, contains the square roots of unity, namely, 1 and -1, but it contains no higher roots of unity (other than 1) because they are all complex.

We begin with a preliminary discussion from group theory.

Lemma 59. *If $C = \langle a \rangle$ is a cyclic group of order n and generator a, then C has a unique subgroup of order d for each divisor d of n, and this subgroup is cyclic.*

Proof. If $n = cd$, we show that a^c has order d (and so $\langle a^c \rangle$ is a subgroup of order d). Clearly $(a^c)^d = a^{cd} = a^n = 1$; we claim that d is the smallest such power. If $(a^c)^r = 1$, then $n \mid cr$ [Theorem G.2(i)]; hence $cr = ns = dcs$ for some integer s, and $r = ds \geq d$.

To prove uniqueness, assume that $\langle x \rangle$ is a subgroup of order d (recall that every subgroup of a cyclic group is cyclic, by Theorem G.1). Now $x = a^m$ and $1 = x^d = a^{md}$; hence $md = nk$ for some integer k. Therefore, $x = a^m = (a^{n/d})^k = (a^c)^k$, so that $\langle x \rangle \subset \langle a^c \rangle$. Since both subgroups have the same order d, it follows that $\langle x \rangle = \langle a^c \rangle$. •

Recall Theorem G.2(ii): If C is a cyclic group with generator x and order n, then x^k is also a generator of C if and only if k and n are relatively prime. It follows that if $g(C)$ denotes the set of all generators of C, and if φ is Euler's function, then

$$|g(C)| = \varphi(n).$$

Theorem 60. *If n is a positive integer, then*

$$n = \sum_{d \mid n} \varphi(d).$$

Proof. Define an equivalence relation on a group G by $x \equiv y$ if $\langle x \rangle = \langle y \rangle$. If $g(C)$ denotes the equivalence class of x, where $C = \langle x \rangle$, then G is the disjoint union

$$G = \bigcup_{C \text{ cyclic}} g(C)$$

If G has order n, then counting gives $n = \sum |g(C)| = \sum \varphi(d)$, where the summation ranges over all cyclic subgroups C of G, while if G is cyclic, then the lemma gives

$$\sum_C |g(C)| = \sum_{d \mid n} \varphi(d),$$

for there is exactly one cyclic subgroup of G for every divisor d of $|G|$. •

Theorem 61. *A group G of order n is cyclic if and only if, for each divisor d of n, there is at most one cyclic subgroup of order d.*

Proof. If G is cyclic, then the result follows from Lemma 59. Conversely, write G as a disjoint union (as in the preceding proof): $G = \bigcup g(C)$. Hence $n = |G| = \sum |g(C)|$, where the summation is over all cyclic subgroups C of G. Since G has at most one cyclic subgroup of order d, Theorem 60 gives

$$n = \sum |g(C)| \leq \sum \varphi(d) = n.$$

Therefore, G has exactly one cyclic subgroup of order d for every divisor d of n; in particular, there is a cyclic subgroup of order n, and so G is cyclic. •

Theorem 62. *If F is a field with multiplicative group $F^\# = F - \{0\}$, then every finite subgroup G of $F^\#$ is cyclic.*

Proof. Suppose that $|G| = n$ and $d \mid n$. If C is a cyclic subgroup of G of order d, then Lagrange's theorem gives $x^d = 1$ for each of the d elements $x \in C$. Were there a second cyclic subgroup of order d, then it would have at least one element not in C, so that G would contain at least $d + 1$ elements x with $x^d = 1$. But the polynomial $x^d - 1$ has at most d roots in a field, and so G has at most one cyclic subgroup of order d. Theorem 61 now shows that G is cyclic. •

Corollary 63. *If n is a fixed positive integer, then all the nth roots of unity in a field F form a cyclic multiplicative group.*

Corollary 64. *If F is a finite field, then $F^\#$ is cyclic and $F = \mathbb{Z}_p(\alpha)$ for some α.*

Proof. If $|F| = q$, take α to be a primitive $(q-1)$st root of unity. •

Remark. Exercise 81 shows that if F is an infinite field, then $F^\#$ is never a cyclic group.

Definition. If F is a finite field of characteristic p, then an element $\alpha \in F$ is called a *primitive element* if $F = \mathbb{Z}_p(\alpha)$.

It follows that any generator of the multiplicative group $F^\#$ is a primitive element.

Let us call an element a in a field F a *square* if there is $u \in F$ with $a = u^2$; i.e., a has a square root in F. In $\mathbb{R}$, an element is a nonsquare if and only if it is negative, and so the product of two nonsquares, being positive, is a square. On the other hand, neither 2 nor 3 is a square in $\mathbb{Q}$, and their product 6 is also not a square. The next corollary shows that, in this respect, finite fields behave more like the reals than the rationals.

Corollary 65. *If F is a finite field and $a, b \in F$ are not squares, then their product ab is a square.*

Proof. If α is a primitive element of F, then every nonzero element of F has the form α^k for some integer k, and α^k is a square if and only if k is even. Since a and b are not squares, we have $a = \alpha^k$ and $b = \alpha^m$ where both k and m are odd. Therefore, $ab = \alpha^{k+m}$ is a square, because $k + m$ is even. •

There is a sophisticated proof of the next example which uses an analysis of Galois groups in algebraic number theory; the following elementary proof is due to G. J. Janusz.

Example 26. The polynomial $f(x) = x^4 - 10x^2 + 1 \in \mathbb{Q}[x]$ is irreducible, but it factors in $\mathbb{Z}_p[x]$ for every prime p.

We know that $f(x)$ is irreducible in $\mathbb{Q}[x]$, by Exercise 67. Completing the square gives

$$x^4 - 10x^2 + 1 = x^4 - 10x^2 + 25 - 24 = (x^2 - 5)^2 - 24.$$

Now regard the coefficients of $f(x)$ as lying in $\mathbb{Z}_p$ for some prime p. If $\sqrt{24} = 2\sqrt{6}$ lies in $\mathbb{Z}_p$, i.e., there is $\beta \in \mathbb{Z}_p$ with $\beta^2 = 6$, then $f(x)$ factors in $\mathbb{Z}_p[x]$ and we are done:[10]

$$f(x) = \left(x^2 - 5 + 2\beta\right)\left(x^2 - 5 - 2\beta\right).$$

We may now assume that the quadratic $x^2 - 6$ has no roots in $\mathbb{Z}_p$, and so $x^2 - 6 \in \mathbb{Z}_p[x]$ is irreducible. By Theorem 45, $E = \mathbb{Z}_p[x]/(x^2 - 6)$ is a field containing an element β with $\beta^2 = 6$, and $\{1, \beta\}$ is a basis of E viewed as a vector space over $\mathbb{Z}_p$.

We claim that $5 + 2\beta$ is a square in E. Every $u \in E$ has a unique expression of the form $u = a + b\beta$, where $a, b \in \mathbb{Z}_p$. If $u^2 = 5 + 2\beta$, then substituting and equating coefficients gives

$$a^2 + 6b^2 = 5 \text{ and } 2ab = 2.$$

We may assume that $p \neq 2$ because $f(x) = x^4 + 1$ does factor in $\mathbb{Z}_2[x]$, so that $ab = 1$ and $b = a^{-1}$. Hence, $a^2 + 6a^{-2} = 5$, which we rewrite as

$$0 = a^4 - 5a^2 + 6 = (a^2 - 2)(a^2 - 3).$$

If neither 2 nor 3 is a square in $\mathbb{Z}_p$, then Corollary 65 would give 6 a square in $\mathbb{Z}_p$, contrary to hypothesis. Therefore, at least one of 2 or 3 is a square.

---

[10]In $\mathbb{Z}_5$, the element $[6] = [1]$ is a square, but $[6]$ is not a square in $\mathbb{Z}_7$, for $[1]$, $[4]$, and $[2] = [9]$ are the only squares in $\mathbb{Z}_7$.

We have shown that there exists $a \in \mathbb{Z}_p$ with $(a + a^{-1}\beta)^2 = 5 + 2\beta$. It now follows that $(a - a^{-1}\beta)^2 = 5 - 2\beta$. Therefore,

$$
\begin{aligned}
f(x) &= (x^2 - 5 + 2\beta)(x^2 - 5 - 2\beta) \\
&= [x^2 - (a - a^{-1}\beta)^2][x^2 - (a + a^{-1}\beta)^2] \\
&= (x + a - a^{-1}\beta)(x - a + a^{-1}\beta)(x + a + a^{-1}\beta)(x - a - a^{-1}\beta) \\
&= (x - a - a^{-1}\beta)(x - a + a^{-1}\beta)(x + a + a^{-1}\beta)(x + a - a^{-1}\beta) \\
&= [(x - a)^2 - a^{-2}\beta^2][(x + a)^2 - a^{-2}\beta^2] \\
&= [(x - a)^2 - 6a^{-2}][(x + a)^2 - 6a^{-2}].
\end{aligned}
$$

We have factored $f(x)$ in $\mathbb{Z}_p[x]$. •

Lemma 66. *If α is a primitive element of $GF(p^n)$, then α is a root of an irreducible polynomial in $\mathbb{Z}_p[x]$ of degree n.*

Proof. If the irreducible polynomial of α over $\mathbb{Z}_p$ has degree d, then $\mathbb{Z}_p(\alpha)$ has order p^d. But this subfield is all of $GF(p^n)$ because α is a primitive element; hence $d = n$. •

It follows from the existence of $GF(p^n)$ that there exist irreducible polynomials in $\mathbb{Z}_p[x]$ of degree n for every $n \geq 1$.

Theorem 67. $\mathrm{Gal}(GF(p^n)/GF(p)) \cong \mathbb{Z}_n$ *with generator $u \mapsto u^p$.*

Remark. This generator is called the *Frobenius automorphism*.

Proof. Denote $GF(p^n)$ by K and denote the Galois group by G. If α is a primitive element, then Lemma 66 says its irreducible polynomial $q(x)$ has degree n, and so K contains at most n of its roots. If $\sigma \in G$, then σ is completely determined by $\sigma(\alpha)$ [because every nonzero element of K has the form α^i and $\sigma(\alpha^i) = \sigma(\alpha)^i$]. But $\sigma(\alpha)$ is a root of $q(x)$ [which has at most $\partial(q) = n$ roots], by Lemma 54; it follows that $|G| \leq n$. On the other hand, $\sigma : u \mapsto u^p$ does lie in G, by Lemma 32; moreover, if $j < n$, then $\sigma^j \neq 1$ (otherwise $u^{p^j} = u$ for all u, and K would contain p^n roots of $x^{p^j} - x$, a contradiction). Thus, σ has order at least n, and so $G = \langle \sigma \rangle$ is cyclic of order n. •

Lemma 68. *Let n be a positive integer and let F be a field. If the characteristic of F is either 0 or is a prime not dividing n, then $x^n - 1$ has n distinct roots in a splitting field.*

Proof. If $f(x) = x^n - 1$, then its derivative $f'(x) = nx^{n-1}$. By hypothesis, this is not zero, and so the gcd $(f, f') = 1$; therefore, $f(x)$ has no repeated roots. •

Lemma 68 fails in a field F of characteristic p when $p \mid n$; for example, $x^p - 1 = (x - 1)^p$ has at most one root (of multiplicity p) in any extension field of F, so that $x^{kp} - 1 = (x^k - 1)^p$ has repeated roots for all $k \geq 1$.

Definition. Let n be a fixed positive integer and let F be a field. A generator of the group of all nth roots of unity is called a *primitive root of unity*.

A primitive nth root of unity in $\mathbb{C}$ is $e^{2\pi i/n}$.

Recall that if R is a ring, then $U(R)$ denotes its multiplicative group of units. In particular,

$$U(\mathbb{Z}_n) = \{ [i] \in \mathbb{Z}_n : (i, n) = 1 \}.$$

When p is a prime, therefore, $U(\mathbb{Z}_p) = \mathbb{Z}_p^{\#}$, the multiplicative group of all nonzero elements.

Theorem 69. *If F is a field and $E = F(\alpha)$, where α is a primitive nth root of unity, then $\mathrm{Gal}(E/F)$ is isomorphic to a subgroup of $U(\mathbb{Z}_n)$, and hence $\mathrm{Gal}(E/F)$ is an abelian group.*

Proof. Since $E = F(\alpha)$, each $\sigma \in \mathrm{Gal}(E/F)$ is completely determined by its value on α. Now $\sigma(\alpha) = \alpha^i$ for some i which is unique mod n, and we denote σ by σ_i, where $0 \leq i \leq n - 1$. Theorem G.2(ii) says that i must be relatively prime to n, for $\sigma | \langle \alpha \rangle$ is an automorphism of $\langle \alpha \rangle$. Therefore, the function $\psi : \sigma_i \mapsto [i]$ is a function $\psi : \mathrm{Gal}(E/F) \to U(\mathbb{Z}_n)$. Now ψ is a homomorphism:

$$\sigma_j \sigma_i(\alpha) = \sigma_j(\alpha^i) = (\alpha^i)^j = \alpha^{ij} = \alpha^{ji}.$$

Hence, $\psi(\sigma_j \sigma_i) = [ji] = \psi(\sigma_j)\psi(\sigma_i)$. This map is injective, by Exercise 73. Therefore $\mathrm{Gal}(E/F)$ is isomorphic to a subgroup of $U(\mathbb{Z}_n)$. •

The multiplicative group $U(\mathbb{Z}_n)$ need not be cyclic; for example, $U(\mathbb{Z}_8)$ consists of the congruence classes [1], [3], [5],[7], and it is isomorphic to the 4-group.

There is a deep partial converse of Theorem 69. The Kronecker–Weber Theorem states that every finite *abelian extension* of $\mathbb{Q}$ [that is, a finite extension $E/\mathbb{Q}$ with $\mathrm{Gal}(E/\mathbb{Q})$ abelian] can be imbedded in a cyclotomic extension $\mathbb{Q}(\omega)$, where ω is some root of unity.

Example 27. If p is a prime, then $\zeta = e^{2\pi i/p}$ is a primitive pth root of unity over $\mathbb{Q}$. As in the proof of Theorem 69, $\mathbb{Q}(\zeta)$ is the splitting field of the cyclotomic polynomial $\Phi_p(x)$ over $\mathbb{Q}$. Since $\Phi_p(x)$ is irreducible over $\mathbb{Q}$, by Corollary 41, we have $|\text{Gal}(\mathbb{Q}(\zeta)/\mathbb{Q})| = p - 1$. The remark after Theorem 69 shows that $\text{Gal}(\mathbb{Q}(\zeta)/\mathbb{Q})$ is isomorphic to a subgroup of $U(\mathbb{Z}_p) = \mathbb{Z}_p^{\#}$. By Corollary 64, the latter group is cyclic, being the multiplicative group of a field. Indeed, $\text{Gal}(\mathbb{Q}(\zeta)/\mathbb{Q}) \cong \mathbb{Z}_p^{\#}$, by Theorem 56.

The homomorphism ψ occurring in Theorem 69 takes values in the multiplicative group $U(\mathbb{Z}_n)$. Here is a variant which takes values in the additive group $\mathbb{Z}_n$.

Theorem 70. *Let F contain a primitive nth root of unity, and let $f(x) = x^n - c \in F[x]$. If E/F is a splitting field of $f(x)$, then there is an injection*

$$\varphi : G = \text{Gal}(E/F) \to \mathbb{Z}_n.$$

Moreover, $f(x)$ is irreducible if and only if φ is surjective.

Proof. If ω is a primitive nth root of unity and if α is a root of $f(x)$, then $\alpha^n = c$, and the list of all the roots of $f(x)$ is $\alpha, \alpha\omega, \dots, \alpha\omega^{n-1}$. If $\sigma \in G$, then $\sigma(\alpha) = \alpha\omega^i$, and σ is completely determined by i; define $\varphi(\sigma) = [i]$ if $\sigma(\alpha) = \alpha\omega^i$. We now show that $\varphi : G \to \mathbb{Z}_n$ is a homomorphism, where $\mathbb{Z}_n$ is the *additive* group. If $\tau \in G$, then $\tau(\omega) = \omega$ (because $\omega \in F$) and $\tau(\alpha) = \alpha\omega^j$ for some j. Hence,

$$\tau\sigma : \alpha \mapsto \alpha\omega^i \mapsto \tau(\alpha\omega^i)$$
$$= \tau(\alpha)\tau(\omega^i)$$
$$= [\alpha\omega^j]\omega^i$$
$$= \alpha\omega^{j+i},$$

so that $\varphi(\tau\sigma) = [j+i] = \varphi(\tau) + \varphi(\sigma)$. Therefore φ is a homomorphism; it is an injection, by Exercise 73. Now φ is surjective if and only if G acts transitively on the roots of $f(x)$. By Exercise 79, this is equivalent to the irreducibility of $f(x)$. $\bullet$

Corollary 71. *Let p be a prime, let F be a field containing a primitive pth root of unity, and let $f(x) = x^p - c \in F[x]$ have splitting field E. Then either $f(x)$ splits and $\text{Gal}(E/F) = 1$ or it is irreducible and $\text{Gal}(E/F) \cong \mathbb{Z}_p$.*

Proof. Consider the map $\mathrm{Gal}(E/F) \to \mathbb{Z}_p$ of the theorem. If $f(x)$ splits, then $\mathrm{Gal}(E/F) = 1$ and its image is trivial; if $f(x)$ does not split, then its image is a nontrivial subgroup of $\mathbb{Z}_p$. But $\mathbb{Z}_p$ has no proper nontrivial subgroups, so that the map must be surjective, $\mathrm{Gal}(E/F) \cong \mathbb{Z}_p$, and $f(x)$ is irreducible. $\bullet$

There is an elementary proof of this corollary that does not assume F contains roots of unity.

Corollary 72. *If p is a prime, F is a field of characteristic 0, and $f(x) = x^p - c \in F[x]$, then either $f(x)$ is irreducible in $F[x]$ or c has a pth root in F.*

Proof.[11] A splitting field E/F of $f(x)$ contains an element α with $\alpha^p = c$ as well as a pth root of unity ω, for if the roots $\alpha, \omega\alpha, \ldots, \omega^{p-1}\alpha$ of $f(x)$ lie in E, then $\omega = (\omega\alpha)/\alpha$ lies in E as well. If $f(x)$ is not irreducible in $F[x]$, then there is a factorization

$$f(x) = g(x)h(x)$$

in $F[x]$ with each factor $g(x)$ and $h(x)$ having degree less than p; let $\partial(g) = k < p$. The constant term b of $g(x)$ is, to sign, the product of some of the roots of $f(x)$ (perhaps with multiplicity), so that $\pm b = \alpha^k \omega^m \in F$ for some integer m. Since ω^m is a pth root of unity and $\alpha^p = c$, we have

$$(\pm b)^p = \alpha^{kp} = c^k.$$

Since p is prime and $k < p$, we have $(k, p) = 1$, and so there are integers s and t with $sk + tp = 1$. Therefore,

$$c = c^{sk+tp} = c^{ks} c^{pt} = (\pm b)^{ps} c^{pt} = [(\pm b)^s c^t]^p,$$

and c has a pth root in F. $\bullet$

Exercises

81. Prove that if F is an infinite field, then its multiplicative group $F^{\#}$ is never cyclic. (Hint. To eliminate the possibility $F^{\#} = \langle u \rangle$, consider the cases of characteristic 0 and characteristic $p > 0$ separately; the latter case should be further subdivided into cases: u transcendental over the prime field $\mathbb{Z}_p$ and u algebraic over $\mathbb{Z}_p$.)

---

[11] The proof works if F has characteristic $q \neq p$.

Solvability by Radicals

We now show how the Galois group takes account of there being a formula for the roots of a polynomial that involves only the field operations and taking square roots, cube roots, etc.

Definition. A field extension B/F is a *pure extension of type m* if $B = F(\alpha)$, where $\alpha^m \in F$ for some positive integer m.

A tower of fields $F = B_0 \subset B_1 \subset \cdots \subset B_t$ is a *radical tower* if each B_{i+1}/B_i is a pure extension. In this case, we call B_t/F a *radical extension of F*.

If $B = F(\alpha)$ is a pure extension of type m and F contains the mth roots of unity, then B/F is a splitting field of $x^m - \alpha^m$.

Definition. If $f(x) \in F[x]$, then $f(x)$ is *solvable by radicals over F* if there is a radical extension B/F which contains a splitting field E of $f(x)$ over F.

We illustrate the definition of solvability by radicals by showing that quadratics, cubics, and quartics over fields of characteristic 0 are solvable by radicals (these formulas are not true for arbitrary fields; for example, the quadratic formula cannot hold when the field of coefficients has characteristic 2).

If $f(x) = x^2 + bx + c \in \mathbb{C}[x]$, define

$$F = \mathbb{Q}(b, c) \text{ and } B = F(\sqrt{b^2 - 4c}).$$

Then B/F is a pure extension of type 2, and B is the splitting field of $f(x)$ over F; therefore, $f(x)$ is solvable by radicals over F.

If $f(x) = x^3 + qx + r \in \mathbb{C}[x]$, define $F = \mathbb{Q}(q, r)$, define

$$B_1 = F(\sqrt{r^2 + 4q^3/27}),$$

which is pure of type 2, and define $B_2 = B_1(y)$, where

$$y^3 = \tfrac{1}{2}(-r + \sqrt{r^2 + 4q^3/27}),$$

which is pure of type 3. The cubic formula says that the roots of $f(x)$ are $y + z$, $\omega y + \omega^2 z$, and $\omega^2 y + \omega z$, where $yz = -q/3$ (so that $z \in B_2$), and ω is a primitive cube roots of unity. Therefore, if we define $B_3 =$

$B_2(\omega)$, which is pure of type 3, then the splitting field E of $f(x)$ is contained in B_3, and $f(x)$ is solvable by radicals. Note that it is possible that E is a proper subfield of B_3, for E need not contain ω; for example, $f(x)$ may have three real roots in which case E is a subfield of $\mathbb{R}$.

If $f(x) = x^4 + qx^2 + rx + s \in \mathbb{C}[x]$, define $F = \mathbb{Q}(q, r, s)$. In the discussion of the quartic formula, we saw that it suffices to find three numbers k, ℓ, and m. Now k^2 is a root of a certain cubic polynomial in $F[x]$, so that there is a radical tower $F \subset B_1 \subset B_2 \subset B_3$ with $k^2 \in B_3$. Define $B_4 = B_3(k)$. Since $2m = k^2 + q + r/k$ and $2\ell = k^2 + q - r/k$, B_4 contains ℓ and m. The quartic formula gives the roots of $f(x)$ as the roots of

$$(x^2 + kx + \ell)(x^2 - kx + m),$$

so that the radical tower can be lengthened two steps, each of type 2, by adjoining $\sqrt{k^2 - 4\ell}$ and $\sqrt{k^2 - 4m}$, with the last extension, B_6, containing the splitting field of $f(x)$. Therefore, $f(x)$ is solvable by radicals.

It should be plain, conversely, that if a polynomial $f(x)$ is solvable by radicals, then there is an expression for its roots in terms of its coefficients, the field operations, and extraction of roots.[12]

Recall that a finite group is *solvable* if it has a normal series with abelian factor groups; moreover Theorem G.20 shows that every quotient and every subgroup of a solvable group is itself solvable.

Lemma 73. *Let F be a field of characteristic 0, let $f(x) \in F[x]$ be solvable by radicals, and let E be a splitting field of $f(x)$ over F.*

(i) *There is a radical tower*

$$F = R_0 \subset R_1 \subset \ldots \subset R_t$$

with $E \subset R_t$, with R_t a splitting field of some polynomial over F, and with each R_i/R_{i-1} is a pure extension of prime type p_i.

(ii) *If R_t/F is a radical extension as in part (i), and if F contains the p_ith roots of unity for all i, then $\mathrm{Gal}(E/F)$ is a solvable group.*

Proof. (i) Since f is solvable by radicals, there is a radical tower

$$F = B_0 \subset B_1 \subset \ldots \subset B_\ell$$

with $E \subset B_\ell$. By Exercise 83, there is an extension K/B_ℓ which is a splitting field of some polynomial in $F[x]$, and by Exercise 85, K/F is

---

[12]According to our definition, the cyclotomic polynomials are solvable by radicals, but it is not so obvious that their roots, namely, the roots of unity, have an expression in terms of radicals; this is, in fact, true, and it is a theorem of Gauss.

also a radical extension. Of course, $E \subset B_\ell \subset K$. Finally, Exercise 82 says that a radical tower from F to K can be refined so that each step is a pure extension of prime type.

(ii) Let

$$F = R_0 \subset R_1 \subset \dots \subset R_t$$

be a radical tower as in part (i), and define

$$G_i = \mathrm{Gal}(R_t/R_i).$$

By hypothesis, F and, hence, each R_i, contains all the p_ith roots of unity, so that each R_i is a splitting field of a polynomial over R_{i-1}. Thus, the hypothesis of Theorem 58 holds and

$$\mathrm{Gal}(R_t/F) = G_0 \supset G_1 \supset \dots \supset G_t = \{1\}$$

is a normal series. The factor groups $\mathrm{Gal}(R_t/R_{i-1})/\mathrm{Gal}(R_t/R_i)$ of this normal series are isomorphic to $\mathrm{Gal}(R_i/R_{i-1})$, by Theorem 58, and these last groups are cyclic of prime order, by Corollary 71. Hence, $\mathrm{Gal}(R_t/F)$ is a solvable group.

Finally, applying Theorem 58 to the tower of fields

$$F \subset E \subset R_t,$$

we see that $\mathrm{Gal}(E/F)$ is a quotient group of the solvable group $\mathrm{Gal}(R_t/F)$, and so it, too, is solvable, by Theorem G.20. •

We now remove the hypothesis that the base field F contain certain roots of unity.

Theorem 74. *Let $f(x) \in F[x]$ be solvable by radicals over a field F of characteristic 0, and let E/F be its splitting field. Then $\mathrm{Gal}(E/F)$ is a solvable group.*

Proof. By hypothesis, there is a radical tower

$$F = R_0 \subset R_1 \subset \dots \subset R_t,$$

with $E \subset R_t$. By Lemma 73(i), we may assume that each R_i/R_{i-1} is of prime type p_i and that R_t/F is a splitting field of some polynomial $h(x) \in F[x]$. Let m be the lcm of the p_i's, and let ω be a primitive mth root of unity. The tower can be lengthened by $R_t \subset R' = R_t(\omega)$, and then refined so that each pure extension in it has prime type. Observe that R' is a splitting field of $(x^m - 1)h(x) \in F[x]$.

Construct a new tower by adjoining ω first:

$$F = R_0 \subset F(\omega) \subset R_1(\omega) \subset \ldots \subset R_t(\omega) = R'.$$

Notice that each extension in this tower is pure and that E, the splitting field of $f(x)$, is contained in R'. Since $F(\omega)/F$ is a splitting field, Theorem 58 gives $\mathrm{Gal}(R'/F(\omega)) \lhd \mathrm{Gal}(R'/F)$ and

$$\mathrm{Gal}(R'/F)/\mathrm{Gal}(R'/F(\omega)) \cong \mathrm{Gal}(F(\omega)/F).$$

Now $\mathrm{Gal}(F(\omega)/F)$ is abelian, hence solvable, by Theorem 69. Each step of the truncated tower

$$F(\omega) \subset R_1(\omega) \subset \ldots \subset R_t(\omega) = R'$$

is a pure extension of prime type, so that Lemma 73(ii), which applies because $F(\omega)$ contains the necessary roots of unity, shows that the normal subgroup $\mathrm{Gal}(R'/F(\omega))$ of $\mathrm{Gal}(R'/F)$ is solvable. Therefore, $\mathrm{Gal}(R'/F)$ is solvable, by Theorem G.21.

Finally, Theorem 58 applies to show that $\mathrm{Gal}(E/F)$ is a quotient group of the solvable group $\mathrm{Gal}(R'/F)$, hence is solvable, by Theorem G.20. •

Of course, this last theorem gives the etymology of the word *solvable* in group theory.

Theorem 75 (Abel–Ruffini). *There exists a quintic polynomial $f(x) \in \mathbb{Q}[x]$ that is not solvable by radicals.*

Proof. If $f(x) = x^5 - 4x + 2$, then $f(x)$ is irreducible over $\mathbb{Q}$, by Eisenstein's criterion. Let $E/\mathbb{Q}$ be the splitting field of $f(x)$ contained in $\mathbb{C}$, and let $G = \mathrm{Gal}(E/\mathbb{Q})$. If α is a root of $f(x)$, then $[\mathbb{Q}(\alpha) : \mathbb{Q}] = 5$, and so

$$[E : \mathbb{Q}] = [E : \mathbb{Q}(\alpha)][\mathbb{Q}(\alpha) : \mathbb{Q}] = 5[E : \mathbb{Q}(\alpha)].$$

By Theorem 56, $|G| = [E : \mathbb{Q}]$ is divisible by 5.

We now use some calculus; $f(x)$ has exactly two critical points, namely, $\pm\sqrt[4]{4/5} \sim \pm.946$, and $f(\sqrt[4]{4/5}) < 0$ and $f(-\sqrt[4]{4/5}) > 0$; it follows easily that $f(x)$ has exactly three real roots (they are, approximately, -1.5185, 0.5085, and 1.2435; the complex roots are $-.1168 \pm 1.4385i$.)

Regarding G as a group of permutations on the 5 roots, we note that G contains a 5-cycle (it contains an element of order 5, by Cauchy's theorem, and the only elements of order 5 in S_5 are 5-cycles). The restriction of complex conjugation, call it σ, is a transposition, for σ interchanges the two complex roots while it fixes the three real roots. By Theorem G.39, S_5 is generated by any transposition and any 5-cycle, so that

$$G = \mathrm{Gal}(E/\mathbb{Q}) \cong S_5$$

is not a solvable group, by Theorem G.34, and Theorem 74 shows that $f(x)$ is not solvable by radicals. •

Remark. Abel and Ruffini proved that the general quintic is not solvable by radicals; that is, there is no formula that works for an arbitrary quintic when one specializes its coefficients (the classical quadratic, cubic, and quartic formulas are of this sort). Theorem 75 is thus a stronger result than what they had shown.

Exercises

82. If E/F is a radical extension over F, then there is a radical tower

$$F = B_0 \subset B_1 \subset \ldots \subset B_t$$

with each B_{i+1}/B_i a pure extension of prime type. (Hint. If $\alpha^n \in F$ and $n = pm$, then there is a tower of fields $F \subset F(\alpha^m) \subset F(\alpha)$.)

83. Let B/F be a finite extension. Prove that there is an extension K/B so that K/F is a splitting field of some polynomial $f(x) \in F[x]$. [13] (Hint. Since B/F is finite, it is algebraic, and there are elements $\alpha_1, \ldots, \alpha_n$ with $B = F(\alpha_1, \ldots, \alpha_n)$. If $p_i(x) \in F[x]$ is the irreducible polynomial of α_i, take K to be a splitting field of $f(x) = p_1(x) \cdots p_n(x)$.)

84. (i) If B and C are subfields of a field E, then their *compositum* $B \vee C$ is the intersection of all the subfields of E containing B and C. Prove that if $\alpha_1, \ldots, \alpha_n \in E$, then $F(\alpha_1) \vee \cdots \vee F(\alpha_n) = F(\alpha_1, \ldots, \alpha_n)$.

(ii) Prove that any splitting field K/F containing B (as in Exercise 83) has the form $K = B_1 \vee \ldots \vee B_r$, where each B_i is isomorphic to B via an isomorphism which fixes F. (Hint: If $\mathrm{Gal}(K/F) = \{\sigma_1, \cdots, \sigma_r\}$, then define $B_i = \sigma_i(B)$.)

85. Using Exercise 84, prove that any splitting field K/F containing a radical extension R_t/F (as in Exercise 83) is itself a radical extension. Conclude that, in the definition of solvable by radicals, one can assume that the last field B_t is a splitting field of some polynomial over F.

---

[13] A smallest such extension K/F is called a *split closure* of B/F; if $f(x)$ is a separable polynomial, then a split closure K/F is called a *normal closure* of B/F.

Independence of Characters

This section introduces the important notion of a fixed field, and characters are used to compute its degree over a base field.

Definition. A *character* of a group G in a field E is a homomorphism $\sigma : G \to E^{\#}$, where $E^{\#} = E - \{0\}$ is the multiplicative group of E.

Definition. A set $\{\sigma_1, \ldots, \sigma_n\}$ of characters of a group G in a field E is *independent*[14] if there do not exist $a_1, \ldots, a_n \in E$, not all 0, with

$$\sum a_i \sigma_i(x) = 0 \quad \text{for all } x \in G.$$

Lemma 76 (Dedekind). *Every set $\{\sigma_1, \ldots, \sigma_n\}$ of distinct characters of a group G in a field E is independent.*

Proof. The proof is by induction on n. If $n = 1$, then $a_1 \sigma_1(x) = 0$ implies that $a_1 = 0$ because $\sigma_1(x) \neq 0$. Let $n > 1$ and assume there is an equation

(1) $\qquad a_1 \sigma_1(x) + \cdots + a_n \sigma_n(x) = 0 \quad \text{for all } x \in G,$

where not all $a_i = 0$. We may assume that every $a_i \neq 0$ lest induction apply; multiplying by a_n^{-1} if necessary, we may further assume that $a_n = 1$. Since $\sigma_n \neq \sigma_1$, there exists $y \in G$ with $\sigma_n(y) \neq \sigma_1(y)$. In Eq. (1), replace x by yx to obtain

$$a_1 \sigma_1(y)\sigma_1(x) + \cdots + a_{n-1}\sigma_{n-1}(y)\sigma_{n-1}(x) + \sigma_n(y)\sigma_n(x) = 0$$

(in this equation, $y \in G$ is fixed while x is an arbitrary element of G). Multiply by $\sigma_n(y)^{-1}$ to obtain an equation

$$a_1 \sigma_n(y)^{-1}\sigma_1(y)\sigma_1(x) + \cdots + \sigma_n(x) = 0;$$

subtract this from Eq. (1) to obtain a sum of $n - 1$ terms

$$a_1[1 - \sigma_n(y)^{-1}\sigma_1(y)]\sigma_1(x) + \cdots = 0.$$

By induction, each of the coefficients is 0. Since $a_1 \neq 0$, we have $1 = \sigma_n(y)^{-1}\sigma_1(y)$; hence $\sigma_n(y) = \sigma_1(y)$, a contradiction.   •

---

[14]All the functions from G to a field E form a vector space $V(G, E)$ over E in which

$$\sigma + \tau : x \mapsto \sigma(x) + \tau(x)$$

and

$$c\sigma : x \mapsto c\sigma(x)$$

for a scalar c. Independence of characters is linear independence in $V(G, E)$.

Corollary 77. *Every set $\{\sigma_1, \ldots, \sigma_n\}$ of distinct automorphisms of a field E is independent.*

Proof. An automorphism σ of E restricts to a (group) homomorphism $\sigma : E^{\#} \to E^{\#}$, hence is a character.   •

Definition. Let $\text{Aut}(E)$ be the group of all the automorphisms of a field E. If G is a subset of $\text{Aut}(E)$, then

$$E^G = \{\alpha \in E : \sigma(\alpha) = \alpha \quad \text{for all } \sigma \in G\}$$

is called the *fixed field*.

It is easy to see that E^G is a subfield of E. The most important instance of this definition is when G is a subgroup of $\text{Aut}(E)$, but there is an application when G is only a subset. Note that

$$H \subset G \quad \text{implies} \quad E^G \subset E^H :$$

if $\alpha \in E$ and $\sigma(\alpha) = \alpha$ for all $\sigma \in G$, then $\sigma(\alpha) = \alpha$ for all $\sigma \in H \subset G$.

Example 28. If E/F is a field extension with Galois group $G = \text{Gal}(E/F)$, then

$$F \subset E^G \subset E;$$

we shall presently consider whether E^G/F is a proper extension.

Example 29. Let $E = F(x_1, \ldots, x_n)$ be the field of rational functions in several variables over a field F. Then $G = S_n$ can be regarded as a subgroup of $\text{Aut}(E)$; it acts by permuting the variables. The elements of the fixed field E^G are called *symmetric functions*[15] over F.

Lemma 78. *If $G = \{\sigma_1, \ldots \sigma_n\}$ is a set of automorphisms of E, then*

$$[E : E^G] \geq n.$$

---

[15] Symmetric functions arise naturally: if

$$f(x) = \prod(x - \alpha_i) = x^n + s_{n-1}x^{n-1} + \cdots + s_1 x + s_0,$$

then each of the coefficients s_j is a symmetric function of the roots $\alpha_1, \ldots, \alpha_n$. This observation is the starting point of Lagrange and Galois (see Appendix D).

Proof. Otherwise $[E : E^G] = r < n$; let $\{\alpha_1, \ldots, \alpha_r\}$ be a basis of E/E^G. Consider the linear system over E of r equations in n unknowns:

$$\sigma_1(\alpha_1)x_1 + \cdots + \sigma_n(\alpha_1)x_n = 0$$
$$\sigma_1(\alpha_2)x_1 + \cdots + \sigma_n(\alpha_2)x_n = 0$$
$$\cdots\cdots\cdots\cdots$$
$$\sigma_1(\alpha_r)x_1 + \cdots + \sigma_n(\alpha_r)x_n = 0.$$

Since $r < n$, there is a nontrivial solution $(x_1, \ldots, x_n)$. For any $\beta \in E$, we have $\beta = \sum b_i \alpha_i$ where $b_i \in E^G$. Multiply the ith row of the system by b_i to obtain the system with ith row:

$$b_i \sigma_1(\alpha_i)x_1 + \cdots + b_i \sigma_n(\alpha_i)x_n = 0.$$

But $b_i = \sigma_j(b_i)$ for all i, j because $b_i \in E^G$. The system thus has ith row:

$$\sigma_1(b_i\alpha_i)x_1 + \cdots + \sigma_n(b_i\alpha_i)x_n = 0.$$

Now add to get
$$\sigma_1(\beta)x_1 + \cdots + \sigma_n(\beta)x_n = 0.$$

The independence of the characters $\{\sigma_1, \ldots, \sigma_n\}$ is violated, for β is an arbitrary element of E. This contradiction proves the theorem. $\bullet$

Theorem 79. *If $G = \{\sigma_1, \ldots, \sigma_n\}$ is a subgroup of $\mathrm{Aut}(E)$, then*

$$[E : E^G] = |G|.$$

Proof. It suffices to prove $[E : E^G] \leq |G|$. Otherwise $[E : E^G] > n$; let $\{\omega_1, \ldots, \omega_{n+1}\}$ be linearly independent vectors in E over E^G. Consider the system of n equations in $n + 1$ unknowns:

$$\sigma_1(\omega_1)x_1 + \cdots + \sigma_1(\omega_{n+1})x_{n+1} = 0$$
$$\cdots\cdots\cdots\cdots$$
$$\sigma_n(\omega_1)x_1 + \cdots + \sigma_n(\omega_{n+1})x_{n+1} = 0.$$

There is a nontrivial solution $(x_1, \ldots, x_{n+1})$ over E; we proceed to normalize it. Choose a solution having the least number r of nonzero components, say $(a_1, \ldots, a_r, 0, \ldots, 0)$; by reindexing the ω_i, we may assume that all nonzero components come first. Note that $r \neq 1$ lest $\sigma_1(\omega_1)a_1 = 0$ imply $a_1 = 0$. Multiplying by its inverse if necessary, we may assume that $a_r = 1$. Not all $a_i \in E^G$ lest the row corresponding to the identity of G violate the linear independence of $\{\omega_1, \ldots, \omega_{n+1}\}$. Our last assumption is that a_1 does not lie in E^G (this, too, can be accomplished by reindexing the ω_i). There thus exists σ_k with $\sigma_k(a_1) \neq a_1$. The original system has jth row

(1) $$\sigma_j(\omega_1)a_1 + \cdots + \sigma_j(\omega_{r-1})a_{r-1} + \sigma_j(\omega_r) = 0.$$

Apply σ_k to this system to obtain

$$\sigma_k\sigma_j(\omega_1)\sigma_k(a_1) + \cdots + \sigma_k\sigma_j(\omega_{r-1})\sigma_k(a_{r-1}) + \sigma_k\sigma_j(\omega_r) = 0.$$

Since G is a group, $\sigma_k\sigma_1, \ldots, \sigma_k\sigma_n$ is just a permutation of $\sigma_1, \ldots, \sigma_n$. Setting $\sigma_k\sigma_j = \sigma_i$, the system has ith row

$$\sigma_i(\omega_1)\sigma_k(a_1) + \cdots + \sigma_i(\omega_{r-1})\sigma_k(a_{r-1}) + \sigma_i(\omega_r) = 0.$$

Subtract this from the ith row of Eq. (1) to obtain a new system with ith row:

$$\sigma_i(\omega_1)[a_1 - \sigma_k(a_1)] + \cdots + \sigma_i(\omega_{r-1})[a_{r-1} - \sigma_k(a_{r-1})] = 0.$$

Since $a_1 - \sigma_k(a_1) \neq 0$, we have found a nontrivial solution of the original system having fewer than r nonzero components, a contradiction.   •

Corollary 80. *If G, H are finite subgroups of* Aut(E) *with $E^G = E^H$, then $G = H$.*

Proof. If $\sigma \in G$, then clearly σ fixes E^G. To prove the converse, suppose σ fixes E^G and $\sigma \notin G$. Then E^G is fixed by the $n + 1$ elements in $G \cup \{\sigma\}$, so that $E^G \subset E^{G \cup \{\sigma\}}$; hence, Lemma 78 and Theorem 79 give the contradiction:

$$n = |G| = [E : E^G] \geq [E : E^{G \cup \{\sigma\}}] \geq n + 1.$$

Therefore, if σ fixes E^G, then $\sigma \in G$.

If $\sigma \in G$, then σ fixes $E^G = E^H$, and hence $\sigma \in H$; the reverse inclusion is proved the same way.   •

Galois Extensions

Our discussion of Galois groups began with a *pair* of fields, namely, an extension E/F that is a splitting field of some polynomial $f(x) \in F[x]$. We are now going to characterize those extension fields of F that are splitting fields of some polynomial in $F[x]$.

Suppose that $G = \text{Gal}(E/F)$; it is easy to see that

$$F \subset E^G \subset E.$$

A natural question is whether $F = E^G$; in general, the answer is no. For example, if $F = \mathbb{Q}$ and $E = \mathbb{Q}(\alpha)$, where α is the real cube root of 2, then $G = \text{Gal}(E/F) = \text{Gal}(\mathbb{Q}(\alpha)/\mathbb{Q}) = \{1\}$ (if $\sigma \in G$, then $\sigma(\alpha)$ is a root of $x^3 - 2$; but E does not contain the other two (complex) roots of this polynomial). Hence $E^G = E \neq F$.

Theorem 81. *The following conditions are equivalent for a finite extension E/F with Galois group $G = \text{Gal}(E/F)$.*

(i) $F = E^G$;

(ii) *every irreducible $p(x) \in F[x]$ with one root in E is separable and has all its roots in E; that is, $p(x)$ splits over E;*

(iii) *E is a splitting field of some separable polynomial $f(x) \in F[x]$.*

Proof. (i) $\Rightarrow$ (ii) Let $p(x) \in F[x]$ be an irreducible polynomial having a root α in E, and let the distinct elements of the set $\{\sigma(\alpha) : \sigma \in G\}$ be $\alpha_1, \ldots, \alpha_n$. Define $g(x) \in E[x]$ by

$$g(x) = \prod(x - \alpha_i).$$

Now each $\sigma \in G$ permutes the α_i; so that each σ fixes each of the coefficients of $g(x)$ (for the coefficients are symmetric functions of the roots); that is, the coefficients of $g(x)$ lie in $E^G = F$. Hence $g(x)$ is a polynomial in $F[x]$ having no repeated roots. Now $p(x)$ and $g(x)$ have a common root in E, and so their gcd in $E[x]$ is not 1; it follows from Corollary 18 that their gcd is not 1 in $F[x]$. Since $p(x)$ is irreducible, it must divide $g(x)$. Therefore $p(x)$ has no repeated roots, hence is separable, and it splits over E.

(ii) $\Rightarrow$ (iii) Choose $\alpha_1 \in E$ with $\alpha_1 \notin F$. Since E/F is a finite extension, α_1 must be algebraic over F; let $p_1(x) \in F[x]$ be its irreducible polynomial. By hypothesis, $p_1(x)$ is a separable polynomial which splits over E; let $K_1 \subset E$ be its splitting field. If $K_1 = E$, we are done. Otherwise, choose $\alpha_2 \in E$ with $\alpha_2 \notin K_1$. By hypothesis, there is a separable irreducible $p_2(x) \in F[x]$ having α_2 as a root. Let $K_2 \subset E$ be the splitting field of $p_1(x)p_2(x)$, a separable polynomial. If $K_2 = E$, we are done; otherwise, iterate this construction. This process must end with $K_m = E$ for some m because E/F is finite.

(iii) $\Rightarrow$ (i) By Theorem 56, $|G| = [E : F]$. But Theorem 79 gives $|G| = [E : E^G]$, so that $[E : F] = [E : E^G]$. Since $F \subset E^G$, it follows that $F = E^G$. •

Definition. A finite field extension E/F is *Galois* (or *normal*) if it satisfies any of the equivalent conditions in Theorem 81.

Remark. Terminology is not yet standard; some authors call Galois extensions *normal*, while others call an extension normal if it is the splitting field of any, not necessarily separable, polynomial.

There is a relative version of Galois extension.

Definition. Given a field extension E/F, an *intermediate field* is a field B with $F \subset B \subset E$.

Definition. Let E/F be a Galois extension and let B and C be intermediate fields. If there exists an isomorphism $B \to C$ fixing F, then C is called a *conjugate* of B.

Theorem 82. *Let E/F be a Galois extension, and let B be an intermediate field. The following conditions are equivalent.*

(i) *B has no conjugates (other than B itself)*;

(ii) *If $\sigma \in \mathrm{Gal}(E/F)$, then $\sigma|B \in \mathrm{Gal}(B/F)$*;

(iii) *B/F is a Galois extension.*

Proof. (i) $\Rightarrow$ (ii) Obvious.

(ii) $\Rightarrow$ (iii) Let $p(x) \in F[x]$ be an irreducible polynomial having a root β in B. Since $B \subset E$ and E/F is Galois, Theorem 81 says that $p(x)$ is a separable polynomial having all its roots in E. Let $\beta' \in E$ be a root of $p(x)$. By Lemma 50, there exists an isomorphism $\tau : F(\beta) \to F(\beta')$ fixing F and with $\tau(\beta) = \beta'$, and τ extends to $\sigma \in \mathrm{Gal}(E/F)$ because E/F is Galois (Theorem 51). By (ii), $\sigma(B) = B$, so that $\beta' = \sigma(\beta) \in \sigma(B) = B$. Therefore, B contains all the roots of $p(x)$, and so $p(x)$ splits in B. Theorem 81 shows that B/F is a Galois extension.

(iii) $\Rightarrow$ (i) B/F is a splitting field of some polynomial $f(x)$ over F, so that $B = F(\alpha_1, \ldots, \alpha_n)$, where $\alpha_1, \ldots, \alpha_n$ are all the roots of $f(x)$. By the proof of Lemma 54, every injective map $\vartheta : B \to E$ fixing F must permute the roots of $f(x)$. It follows that

$$\vartheta(B) = \vartheta(F(\alpha_1, \ldots, \alpha_n)) = F(\vartheta\alpha_1, \ldots, \vartheta\alpha_n) = B. \quad \bullet$$

Example 30. Consider $f(x) = x^3 - 2 \in \mathbb{Q}[x]$. As we have seen in Example 24, a splitting field for $f(x)$ is $E = \mathbb{Q}(\alpha, \omega)$, where $\alpha = \sqrt[3]{2}$ and $\omega = e^{2\pi i/3}$. Since $E/\mathbb{Q}$ is a splitting field of a separable polynomial ($\mathbb{Q}$ is a perfect field), $E/\mathbb{Q}$ is a Galois extension.

If $g(x) = x^3 - 3x^2 + 3x - 3$, then $g(x)$ is irreducible in $\mathbb{Q}[x]$, by Eisenstein's criterion, but it has a root in E, namely, $\beta = 1 + \alpha$. It follows that $g(x)$ splits in $E[x]$, as the reader may check.

The intermediate field $B = \mathbb{Q}(\omega)$ is a Galois extension over $\mathbb{Q}$, for it is a splitting field of $x^3 - 1$. We have seen in Example 24 that $\text{Gal}(E/\mathbb{Q}) \cong S_3$. It follows that $\sigma(B) = B$ for every $\sigma \in \text{Gal}(E/\mathbb{Q})$. On the other hand, if $C = \mathbb{Q}(\alpha)$, then $\mathbb{Q}(\alpha^2)$ is a conjugate of C, and $\mathbb{Q}(\alpha^2) \neq C$.

Exercises

86. If E/F is a Galois extension and B is an intermediate field, then E/B is a Galois extension.

87. If F has characteristic $\neq 2$ and E/F is a field extension with $[E : F] = 2$, then E/F is Galois.

88. Show that being Galois need not be transitive; that is, if $F \subset B \subset E$ and E/B and B/F are Galois, then E/F need not be Galois. (Hint: Consider $\mathbb{Q} \subset \mathbb{Q}(\alpha) \subset \mathbb{Q}(\beta)$, where α is a square root of 2 and β is a fourth root of 2.)

89. Let $E = F(x_1, \ldots, x_n)$ and let S be the subfield of all symmetric functions. Prove that $[E : S] = n!$ and $\text{Gal}(E/S) \cong S_n$. (Hint: Show that E/S is a splitting field of the separable polynomial $f(t) = \prod(t - x_i)$.)

90. Let E/F be a Galois extension and let $p(x) \in F[x]$ be irreducible. Show that all the irreducible factors of $p(x)$ in $E[x]$ have the same degree. (Hint: Use Exercise 84.)

91. Given a field F and a finite group G of order n, show that there is a subfield $K \subset E = F(x_1, \ldots, x_n)$ with $\text{Gal}(E/K) \cong G$. (Hint: Use Exercise 89 and Cayley's theorem (Theorem G.24).)

The Fundamental Theorem of Galois Theory

Given a Galois extension E/F, the fundamental theorem will show a strong connection between the subgroups of $\mathrm{Gal}(E/F)$ and the intermediate fields between F and E.

Definition. A *lattice* is a partially ordered set $(L, \preceq)$ in which each pair of elements $a, b \in L$ has a least upper bound $a \vee b$ and a greatest lower bound $a \wedge b$.

Recall that a nonempty set L is a *partially ordered set* if $\preceq$ is a reflexive, transitive, and antisymmetric binary relation on L. An element c is an *upper bound* of a and b if $a \preceq c$ and $b \preceq c$; an element d is a *least upper bound* of a and b if it is an upper bound with $d \preceq c$ for every upper bound c. Greatest lower bound is defined analogously, reversing the inequalities.

Example 31. If X is a set, let L be the family of all the subsets of X, and define $A \preceq B$ to mean $A \subset B$. Then L is a lattice with

$$A \vee B = A \cup B \quad \text{and} \quad A \wedge B = A \cap B.$$

Example 32. If G is a group, let $\mathrm{Sub}(G)$ be the family of all the subgroups of G, and define $H \preceq K$ to mean $H \subset K$. Then $\mathrm{Sub}(G)$ is a lattice with $H \vee K$ the subgroup generated by H and K, and $H \wedge K = H \cap K$.

Example 33. Let E/F be a field extension, let $\mathrm{Lat}(E/F)$ be the family of all intermediate fields, and define $B \preceq C$ to mean $B \subset C$. Then $\mathrm{Lat}(E/F)$ is a lattice with $B \vee C$ their compositum and $B \wedge C = B \cap C$.

Example 34. Let L be the set of all integers $n \geq 1$, and define $n \preceq m$ to mean $n \mid m$. Then L is a lattice with $n \vee m = \mathrm{lcm}\{n, m\}$ and $n \wedge m = \gcd\{n, m\}$.

The next result generalizes the De Morgan laws, where $L = L' = P(X)$, the power set of a set X, and γ is complementing.

Lemma 83. *If L and L' are lattices and $\gamma : L \to L'$ is a bijection such that both γ and γ^{-1} are **order reversing** $[a \preceq b$ implies $\gamma(b) \preceq \gamma(a)]$, then*

$$\gamma(a \vee b) = \gamma(a) \wedge \gamma(b) \quad \text{and} \quad \gamma(a \wedge b) = \gamma(a) \vee \gamma(b).$$

Proof. Now $a, b \preceq a \vee b$ implies $\gamma(a), \gamma(b) \succeq \gamma(a \vee b)$; that is, $\gamma(a \vee b)$ is a lower bound of $\gamma(a), \gamma(b)$. It follows that $\gamma(a) \wedge \gamma(b) \succeq \gamma(a \vee b)$; since γ is surjective, there is $c \in L$ with $\gamma(a) \wedge \gamma(b) = \gamma(c)$. Apply γ^{-1} (which is order reversing, by hypothesis) to obtain $a, b \preceq c \preceq a \vee b$. Hence $c = a \vee b$ and $\gamma(a \vee b) = \gamma(c) = \gamma(a) \wedge \gamma(b)$. A similar argument proves the other half of the statement. •

Theorem 84 (Fundamental Theorem of Galois Theory). *Let E/F be a Galois extension with Galois group $G = \mathrm{Gal}(E/F)$.*

(i) *The function $\gamma : \mathrm{Sub}(G) \to \mathrm{Lat}(E/F)$, defined by $H \mapsto E^H$, is an order reversing bijection with inverse $\delta : B \mapsto \mathrm{Gal}(E/B)$.*

(ii) *$E^{\mathrm{Gal}(E/B)} = B$ and $\mathrm{Gal}(E/E^H) = H$.*

(iii)
$$E^{H \vee K} = E^H \cap E^K;$$
$$E^{H \cap K} = E^H \vee E^K;$$
$$\mathrm{Gal}(E/B \vee C) = \mathrm{Gal}(E/B) \cap \mathrm{Gal}(E/C);$$
$$\mathrm{Gal}(E/B \cap C) = \mathrm{Gal}(E/B) \vee \mathrm{Gal}(E/C).$$

(iv) *$[B : F] = [G : \mathrm{Gal}(E/B)]$ and $[G : H] = [E^H : F]$.*

(v) *B/F is a Galois extension if and only if $\mathrm{Gal}(E/B)$ is a normal subgroup of G.*

Proof. (i) It is easy to see that γ is order reversing: $K \leq H$ implies $E^H \leq E^K$. That γ is injective is precisely the statement of Corollary 80. To see that γ is surjective, consider the composite

$$\mathrm{Lat}(E/F) \xrightarrow{\delta} \mathrm{Sub}(G) \xrightarrow{\gamma} \mathrm{Lat}(E/F),$$

where δ is the map $B \mapsto \mathrm{Gal}(E/B)$. Then $\gamma\delta : B \mapsto \mathrm{Gal}(E/B) \mapsto E^{\mathrm{Gal}(E/B)}$. By Exercise 83, E/F Galois implies that E/B is Galois for every intermediate field B; hence Theorem 81 gives $B = E^{\mathrm{Gal}(E/B)}$; hence $\gamma\delta$ is the identity and γ is a surjection. It follows that γ is a bijection with inverse δ.

(ii) This is just the statement that $\delta\gamma$ and $\gamma\delta$ are identity functions.

(iii) The first pair of equations follows from Lemma 83 because γ is an order reversing bijection; the second pair follows because $\delta = \gamma^{-1}$ is also an order reversing bijection.

(iv) $[B : F] = [E : F]/[E : B] = |G|/|\mathrm{Gal}(E/B)| = [G : \mathrm{Gal}(E/B)],$

so that the degree of B/F is the index of $\text{Gal}(E/B)$ in G. The second equation follows from setting $B = E^H$, because $\text{Gal}(E/E^H) = H$.

(v) If B/F is Galois, then we have seen, in Theorem 58, that $\text{Gal}(E/B)$ is a normal subgroup of G. Conversely, suppose that H is a normal subgroup of G; is E^H/F a Galois extension? If $\sigma \in G$, $\tau \in H$, and $\alpha \in E^H$, then $\tau\sigma(\alpha) = \sigma\tau'(\alpha)$ for some $\tau' \in H$, by normality of H in G, and $\sigma\tau'(\alpha) = \sigma(\alpha)$ because τ' fixes α. Therefore $\alpha \in E^H$ implies $\sigma(\alpha) \in E^H$; that is, $\sigma(E^H) \subset E^H$; indeed, $\sigma(E^H) = E^H$ because both have the same dimension over F. By Theorem 82, E^H/F is a Galois extension.   •

Applications

Corollary 85. *A Galois extension E/F has only finitely many intermediate fields.*

Proof. The Galois group $\text{Gal}(E/F)$, being finite, has only finitely many subgroups.   •

Theorem 86 (Steinitz). *A finite extension E/F is simple if and only if it has only finitely many intermediate fields.*

Proof. Assume that $E = F(\alpha)$ and let $p(x)$ be the irreducible polynomial of α over F. Given an intermediate field B, let $g(x)$ be the irreducible polynomial of α over B. If B' is the subfield of B generated by F and the coefficients of $g(x)$, then $g(x)$ is also irreducible over B'. Since $E = B(\alpha) = B'(\alpha)$, it follows that $[E : B] = [B(\alpha) : B]$ and $[E : B'] = [B'(\alpha) : B']$; hence $[E : B] = [E : B']$, for both equal the degree of $g(x)$. Therefore, $B = B'$ and B is completely determined by $g(x)$. But $g(x)$ is a divisor of $p(x)$; as there are only finitely many monic divisors of $p(x)$ over E, there are only finitely many intermediate fields B.

Assume that E/F has only finitely many intermediate fields. If F is finite, then Corollary 64 shows that E/F is simple: just adjoin a primitive element. We may, therefore, assume that F is infinite. Now $E = F(\alpha_1, \dots, \alpha_n)$; by induction on n, it suffices to prove that $E = F(\alpha, \beta)$ is a simple extension. Consider all elements γ of the form $\gamma = \alpha + t\beta$, where $t \in F$; there are infinitely many such γ because F is infinite. Since there are only finitely many intermediate fields, there are only finitely many fields of the form $F(\gamma)$. There thus exist distinct elements $t, t' \in F$ with

$F(\gamma) = F(\gamma')$, where $\gamma' = \alpha + t'\beta$. Clearly, $F(\gamma) \subset F(\alpha, \beta)$. For the reverse inclusion, $F(\gamma) = F(\gamma')$ contains $\gamma - \gamma' = (t - t')\beta$. Since $t \neq t'$, we have $\beta \in F(\gamma)$. But now $\alpha = \gamma - t\beta \in F(\gamma)$, so that $F(\alpha, \beta) \subset F(\gamma)$, as desired. •

Corollary 87. *If E/F is a finite simple extension and B is an intermediate field, then B/F is simple.*

Corollary 88 **(Theorem of the Primitive Element).** *Every finite separable extension B/F is simple.*

Proof. By Exercise 83, there is a Galois extension E/F having B as an intermediate field. By Corollary 85, E/F has only finitely many intermediate fields. Therefore, B/F has only finitely many intermediate fields, and so B/F is simple, by Theorem 86. •

Using the proof of Theorem 86, it is easy to show that one may choose a primitive element of $F(\alpha_1, \ldots, \alpha_n)$ of the form $t_1\alpha_1 + \cdots + t_n\alpha_n$ for $t_i \in F$.

Corollary 89. *The Galois field $GF(p^n)$ has exactly one subfield of order p^d for every divisor d of n.*

Proof. We have seen in Theorem 67 that $\mathrm{Gal}(GF(p^n)/GF(p)) \cong \mathbb{Z}_n$; moreover, Lemma 59 shows that a cyclic group of order n has exactly one subgroup of order d for every divisor d of n. Now a subgroup of order d has index n/d, and so the Fundamental Theorem says that the corresponding intermediate field has degree n/d. But the numbers n/d, as d varies over all the divisors of n, themselves vary over all the divisors of n. •

Even more is true. The lattice of all intermediate fields is the same as the lattice of all subgroups of $\mathbb{Z}_n$, and this, in turn, is the same as the lattice of all the divisors of n under lcm and gcd (a sublattice of the lattice of Example 34).

Corollary 90. *If E/F is an abelian extension, i.e., a Galois extension whose Galois group $\mathrm{Gal}(E/F)$ is abelian, then every intermediate field B/F is a Galois extension.*

Proof. Every subgroup of an abelian group is a normal subgroup. •

Corollary 91. *Let $f(x) \in F[x]$ be a separable polynomial, and let E/F be a splitting field. Let $f(x) = g(x)h(x)$ in $F[x]$, and let B/F and C/F*

*be splitting fields of $g(x)$, $h(x)$, respectively, contained in E. If $B \cap C = F$ (such fields are called **linearly disjoint** over F), then*

$$\mathrm{Gal}(E/F) \cong \mathrm{Gal}(B/F) \times \mathrm{Gal}(C/F).$$

Proof. Recall that if H and K are subgroups of a group G, then G is their *direct product*, denoted by $G = H \times K$, if both H and K are normal subgroups, $H \cap K = \{1\}$, and $H \vee K = G$. Since B/F and C/F are Galois extensions, both $\mathrm{Gal}(E/B)$ and $\mathrm{Gal}(E/C)$ are normal subgroups of $\mathrm{Gal}(E/F)$. The hypothesis gives $B \vee C = E$, so that

$$\mathrm{Gal}(E/B) \cap \mathrm{Gal}(E/C) = \mathrm{Gal}(E/B \vee C) = \mathrm{Gal}(E/E) = \{1\}.$$

Also, linear disjointness gives

$$\mathrm{Gal}(E/B) \vee \mathrm{Gal}(E/C) = \mathrm{Gal}(E/B \cap C) = \mathrm{Gal}(E/F).$$

Hence $\mathrm{Gal}(E/F)$ is a direct product:

$$\mathrm{Gal}(E/F) = \mathrm{Gal}(E/B) \times \mathrm{Gal}(E/C).$$

Finally, a general fact about arbitrary groups H and K, namely,

$$(H \times K)/H \cong K,$$

gives

$$\mathrm{Gal}(E/F)/\mathrm{Gal}(E/B) \cong \mathrm{Gal}(E/C),$$

while Theorem 58 gives $\mathrm{Gal}(E/F)/\mathrm{Gal}(E/B) \cong \mathrm{Gal}(B/F)$. Therefore,

$$\mathrm{Gal}(E/C) \cong \mathrm{Gal}(B/F).$$

Similarly, $\mathrm{Gal}(E/B) \cong \mathrm{Gal}(C/F)$, as desired.   •

The fundamental theorem can also suggest counterexamples, for it translates problems about fields (which are usually infinite structures) into problems about finite groups. For example, let E/F be a Galois extension, and let B and C be intermediate fields of degree 2^b and 2^c, respectively; is the degree of their compositum $B \vee C$ also a power of 2? If $G = \mathrm{Gal}(E/F)$ and H and K are the subgroups corresponding to B and C, respectively, then the fundamental theorem gives

$$[B \vee C : F] = [G : H \cap K].$$

The translated question is: If both $[G : H]$ and $[G : K]$ are powers of 2, must $[G : H \cap K]$ be a power of 2? In Exercise 89, we saw that there is a Galois extension E/F with $\mathrm{Gal}(E/F) \cong S_4$. Let H be the subgroup of all permutations of $\{1, 2, 3\}$ (that is, all $\sigma \in S_4$ with $\sigma(4) = 4$) and let K be the subgroup of all permutations of $\{2, 3, 4\}$. Now $[S_4 : H] = 4 = [S_4 : K]$, but $[S_4 : H \cap K] = 12$ (because $H \cap K = \{(1), (23)\}$ has order 2).

We are now going to prove the fundamental theorem of algebra, first proved by Gauss (1799). Assume that $\mathbb{R}$ satisfies a weak form of the intermediate value theorem: if $f(x) \in \mathbb{R}[x]$ and there exist $a, b \in \mathbb{R}$ such that $f(a) > 0$ and $f(b) < 0$, then $f(x)$ has a real root. Here are some preliminary consequences.

(1) *Every positive real r has a real square root.*

If $f(x) = x^2 - r$, then $f(1 + r) > 0$ and $f(0) < 0$.

(2) *Every quadratic $g(x) \in \mathbb{C}[x]$ has a complex root.*

First, every complex number z has a complex square root. Write z in polar form: $z = re^{i\theta}$, where $r \geq 0$, and $\sqrt{z} = \sqrt{r}e^{i\theta/2}$. It follows that the quadratic formula can give the (complex) roots of $g(x)$.

(3) *The field $\mathbb{C}$ has no extensions of degree 2.*

Such an extension would contain an element whose irreducible polynomial is a quadratic in $\mathbb{C}[x]$, and (2) shows that no such polynomial exists.

(4) *Every $f(x) \in \mathbb{R}[x]$ having odd degree has a real root.*

Let $f(x) = a_0 + a_1 x + \ldots + a_{n-1} x^{n-1} + x^n \in \mathbb{R}[x]$. Define $t = 1 + \sum |a_i|$. Now $|a_i| \leq t - 1$ for all i, and

$$
\begin{aligned}
|a_0 + a_1 t + \ldots + a_{n-1} t^{n-1}| &\leq (t - 1)[1 + t + \ldots + t^{n-1}] \\
&= t^n - 1 \\
&< t^n.
\end{aligned}
$$

It follows that $f(t) > 0$ (for any not necessarily odd n) because the sum of the early terms is dominated by t^n. When n is odd, $f(-t) < 0$, for

$$(-t)^n = (-1)^n t^n < 0,$$

and so the same estimate as above now shows that $f(-t) < 0$.

(5) *There is no field extension $E/\mathbb{R}$ of odd degree > 1.*

If $\alpha \in E$, then its irreducible polynomial must have even degree, by (4), so that $[\mathbb{R}(\alpha) : \mathbb{R}]$ is even. Hence $[E : \mathbb{R}] = [E : \mathbb{R}(\alpha)][\mathbb{R}(\alpha) : \mathbb{R}]$ is even.

Theorem 92 (Fundamental Theorem of Algebra). *Every nonconstant $f(x) \in \mathbb{C}[x]$ has a complex root.*

Proof. If $f(x) = \sum a_i x^i \in \mathbb{C}[x]$, define $\overline{f}(x) = \sum \overline{a}_i x^i$, where $\overline{a}_i$ is the complex conjugate of a_i. If $f(x)\overline{f}(x) = \sum c_k x^k$, then $c_k = \sum_{i+j=k} a_i \overline{a}_j$; it follows that $\overline{c}_k = c_k$, so that $f(x)\overline{f}(x) \in \mathbb{R}[x]$. Since $f(x)$ has a complex root if and only if $f(x)\overline{f}(x)$ has a complex root, it suffices to prove that every real polynomial has a complex root.

Let $p(x)$ be an irreducible polynomial in $\mathbb{R}[x]$, and let $E/\mathbb{R}$ be a splitting field of $(x^2 + 1)p(x)$ which contains $\mathbb{C}$. Since $\mathbb{R}$ has characteristic 0, $E/\mathbb{R}$ is a Galois extension; let G be its Galois group. If $|G| = 2^m k$, where k is odd, then G has a subgroup H of order 2^m, by the Sylow theorem (Theorem G.13); let $B = E^H$ be the corresponding intermediate field. Now the degree $[B : \mathbb{R}]$ equals the index $[G : H] = k$. But we have seen above that $\mathbb{R}$ has no extension of odd degree > 1; hence $k = 1$ and G is a 2-group. By Theorem G.23, the subgroup $\mathrm{Gal}(E/\mathbb{C})$ of G (corresponding to $\mathbb{C}$) has a subgroup of index 2 provided $|\mathrm{Gal}(E/\mathbb{C})| > 1$; its corresponding intermediate field is an extension of $\mathbb{C}$ of degree 2, and this contradicts (3) above. We conclude that $\mathrm{Gal}(E/\mathbb{C}) = \{1\}$ and $E = \mathbb{C}$. •

Corollary 93. *Every $f(x) \in \mathbb{C}[x]$ of degree $n \geq 1$ splits over $\mathbb{C}$; that is, $f(x)$ has a factorization*

$$f(x) = c(x - \alpha_1) \cdots (x - \alpha_n),$$

where $c, \alpha_1, \ldots, \alpha_n \in \mathbb{C}$.

Proof. An easy induction on $n \geq 1$. •

Remark. A field K is called ***algebraically closed*** if every $f(x) \in K[x]$ has a root in K (thus, $\mathbb{C}$ is algebraically closed). It can be proved that every field F has an algebraically closed extension; indeed, it has a smallest such, which is called its ***algebraic closure***.

Exercises

92. Let E/F be a Galois extension with $[E : F] > 1$.

    (i) Must there be an intermediate field of prime degree over F? (Hint: The alternating group A_6 has no subgroups of prime index [see Theorem G.37].)

    (ii) Same question as in (i) with the added hypothesis that $\text{Gal}(E/F)$ is a solvable group.

93. Show that $\mathbb{Z}_p(x, y)$ is a finite extension of its subfield $\mathbb{Z}_p(x^p, y^p)$, but it is not a simple extension.

94. Let $K = \mathbb{Z}_p(t)$ be the field of rational functions, let $f(x) = x^p - x - t \in K[x]$, and let E/K be a splitting field of $f(x)$. Prove that $\text{Gal}(E/K) \cong \mathbb{Z}_p$ but that $f(x)$ is not solvable by radicals.

Galois's Great Theorem

We prove the converse of Theorem 74: Solvability of the Galois group of $f(x) \in F[x]$, where F is a field of characteristic 0, implies $f(x)$ is solvable by radicals. We begin with some lemmas; the first one has a quaint name signifying its use as a device to get around the possible absence of roots of unity in the ground field.

Lemma 94 (Accessory Irrationalities). *Let E/F be a splitting field of $f(x) \in F[x]$ with Galois group $G = \text{Gal}(E/F)$. If F^*/F is an extension and E^*/F^* is a splitting field of $f(x)$ containing E, then restriction $\sigma \mapsto \sigma|E$ is an injective homomorphism*

$$\text{Gal}(E^*/F^*) \to \text{Gal}(E/F).$$

Proof. The hypothesis gives

$$E = F(\alpha_1, \dots, \alpha_n) \text{ and } E^* = F^*(\alpha_1, \dots, \alpha_n),$$

where $\alpha_1, \dots, \alpha_n$ are the roots of $f(x)$. If $\sigma \in \text{Gal}(E^*/F^*)$, then σ permutes the α_i's and fixes F^*, hence F; therefore, $\sigma|E \in \text{Gal}(E/F)$. Using Exercise 73, one sees that $\sigma \mapsto \sigma|E$ is an injection. •

Definition. If E/F is a Galois extension and $\alpha \in E^{\#} = E - \{0\}$, define its *norm* $N(\alpha)$ by

$$N(\alpha) = \prod_{\sigma \in \mathrm{Gal}(E/F)} \sigma(\alpha).$$

Here are some preliminary properties of the norm whose simple proofs are left as exercises. In (i) and (iv), $G = \mathrm{Gal}(E/F)$.

(i) If $\alpha \in E^{\#}$, then $N(\alpha) \in F^{\#}$ (because $N(\alpha) \in E^G = F$).

(ii) $N(\alpha\beta) = N(\alpha)N(\beta)$, so that $N : E^{\#} \to F^{\#}$ is a homomorphism.

(iii) If $a \in F^{\#}$, then $N(a) = a^n$, where $n = [E : F]$.

(iv) If $\sigma \in G$ and $\alpha \in E^{\#}$, then $N(\sigma(\alpha)) = N(\alpha)$.

Given a homomorphism, one asks about its kernel and image. The image of the norm is not easy to compute; the next result (which was the ninetieth theorem in an exposition of Hilbert (1897) on algebraic number theory) computes the kernel of the norm in a special case.

Lemma 95 (Hilbert's Theorem 90). *Let E/F be a Galois extension whose Galois group $G = \mathrm{Gal}(E/F)$ is cyclic of order n; let σ be a generator of G. Then $N(\alpha) = 1$ if and only if there exists $\beta \in E^{\#}$ with*

$$\alpha = \beta\sigma(\beta)^{-1}.$$

Proof. If $\alpha = \beta\sigma(\beta)^{-1}$, then

$$
\begin{aligned}
N(\alpha) &= N(\beta\sigma(\beta)^{-1}) = N(\beta)N(\sigma(\beta)^{-1}) \\
&= N(\beta)N(\sigma(\beta))^{-1} = N(\beta)N(\beta)^{-1} = 1.
\end{aligned}
$$

For the converse, define "partial norms":

$$
\begin{aligned}
\delta_0 &= \alpha, \\
\delta_1 &= \alpha\sigma(\alpha), \\
\delta_2 &= \alpha\sigma(\alpha)\sigma^2(\alpha), \\
&\vdots \\
\delta_{n-1} &= \alpha\sigma(\alpha)\cdots\sigma^{n-1}(\alpha) = N(\alpha) = 1.
\end{aligned}
$$

It is easy to see that

(1) $$\alpha\sigma(\delta_i) = \delta_{i+1} \quad \text{for all } 0 \le i \le n - 2.$$

By independence of the characters $\{1, \sigma, \sigma^2, \ldots, \sigma^{n-1}\}$, there exists $\gamma \in E$ with

$$\delta_0 \gamma + \delta_1 \sigma(\gamma) + \cdots + \delta_i \sigma^i(\gamma) + \cdots + \delta_{n-2} \sigma^{n-2}(\gamma) + \sigma^{n-1}(\gamma) \neq 0;$$

call this sum β. Using Eq. (1), one easily checks that

$$\sigma(\beta) = \alpha^{-1}[\delta_1 \sigma(\gamma) + \cdots + \delta_i \sigma^i(\gamma) + \cdots + \delta_{n-1} \sigma^{n-1}(\gamma)] + \sigma^n(\gamma).$$

But $\sigma^n = 1$, so that the last term is just $\gamma = \alpha^{-1} \delta_0 \gamma$. Hence $\sigma(\beta) = \alpha^{-1}\beta$, as desired.   •

Corollary 96. *Let E/F be a Galois extension of prime degree p. If F has a primitive pth root of unity, then $E = F(\beta)$, where $\beta^p \in F$, and so E/F is a pure extension.*

Proof. If ω is a primitive pth root of unity, then $N(\omega) = \omega^p = 1$, because $\omega \in F$. Now $G = \mathrm{Gal}(E/F) \cong \mathbb{Z}_p$, by Corollary 71, hence is cyclic; let σ be a generator. By Hilbert's Theorem 90, we have $\omega = \beta\sigma(\beta)^{-1}$ for some $\beta \in E$. Hence $\sigma(\beta) = \beta\omega^{-1}$. It follows that $\sigma(\beta^p) = (\beta\omega^{-1})^p = \beta^p$, and so $\beta^p \in E^G = F$ because σ generates G and E/F is Galois. Note that $\beta \notin F$, lest $\omega = 1$, so that $F(\beta) \neq F$ is an intermediate field. Therefore $E = F(\beta)$, because $[E : F] = p$, and hence E has no proper intermediate fields.   •

Here is another proof of this last corollary that uses neither Hilbert's Theorem 90 nor the norm. The existence of an element $\beta \in E$ with $\omega = \beta\sigma(\beta)^{-1}$ is shown by an elegant application of linear algebra.[16] (We have given the first proof because the norm is a very important tool in algebraic number theory, and Hilbert's Theorem 90 itself is a useful result that is one of the early theorems involving homological algebra.)

Corollary 97. *Let E/F be a Galois extension with $\mathrm{Gal}(E/F) = \langle\sigma\rangle \cong \mathbb{Z}_p$, where p is a prime. If F contains a primitive pth root of unity ω, then there is an element $\beta \in E$ with $\omega = \beta\sigma(\beta)^{-1}$.*

Proof. View E as a vector space over F and $\sigma : E \to E$ as a linear transformation. Since $\sigma^p = 1$, we see that σ satisfies the polynomial $x^p - 1$. Now σ satisfies no polynomial of smaller degree, lest we contradict independence of the characters $1, \sigma, \ldots, \sigma^{p-1}$. Therefore, $x^p - 1$ is the

---

[16]E. Houston, *A Linear Algebra Approach to Cyclic Extensions in Galois Theory*, Amer. Math. Monthly 100 (1993), 64–66.

minimal polynomial of σ; indeed, $x^p - 1$ is the characteristic polynomial of σ. Since ω^{-1} is a root of $x^p - 1$, it is an eigenvalue of σ (remember that $\omega^{-1} \in F$). If β is an eigenvector of σ corresponding to ω^{-1}, then $\sigma(\beta) = \omega^{-1}\beta$. Therefore, $\omega = \beta\sigma(\beta)^{-1}$. •

Theorem 98 (Galois). *Let F be a field of characteristic 0, and let E/F be a Galois extension. Then $G = \mathrm{Gal}(E/F)$ is a solvable group if and only if E can be imbedded in a radical extension of F.*

Therefore, the Galois group of $f(x) \in F[x]$, where F is a field of characteristic 0, is a solvable group if and only if $f(x)$ is solvable by radicals.

Proof. Sufficiency is Theorem 74, and we now prove the converse. Since G is solvable, Corollary G.17 provides a normal subgroup H of prime index, say, p. Let ω be a primitive pth root of unity, which exists because F has characteristic 0. We first prove the theorem, by induction on $[E : F]$, assuming that $\omega \in F$. The base step is obviously true. For the inductive step, consider the intermediate field E^H. Now E/E^H is a Galois extension (by Exercise 86), $\mathrm{Gal}(E/E^H)$ is a solvable group (being a subgroup of the solvable group $\mathrm{Gal}(E/F) = G$), and $[E : E^H] < [E : F]$. By induction, there is a radical tower

$$E^H \subset R_1 \subset \cdots \subset R_m,$$

where $E \subset R_m$. Now E^H/F is a Galois extension, because $H \lhd G$, having degree $[E^H : F] = [G : H] = p$. Since we are assuming that F contains ω, Corollary 96 gives $E^H = F(\beta)$, where $\beta^p \in F$; that is, E^H/F is a pure extension. Hence, the radical tower can be lengthened by adding the prefix $F \subset E^H$, thereby displaying R_m/F as a radical extension.

For the general case, define $F^* = F(\omega)$ and $E^* = E(\omega)$. Observe that E^*/F is a Galois extension, for if E/F is a splitting field of $f(x) \in F[x]$, then E^*/F is a splitting field of the necessarily separable polynomial $f(x)(x^p - 1)$. It follows that E^*/F^* is also a Galois extension; let $G^* = \mathrm{Gal}(E^*/F^*)$. By Lemma 94, there is an injection $G^* \to G = \mathrm{Gal}(E/F)$, so that G^* is solvable (being isomorphic to a subgroup of a solvable group). Since $\omega \in F^*$, we know that E^*, and hence its subfield E, can be imbedded in a radical extension R^*/F^*; there is a radical tower

$$F^* \subset R_1^* \subset \cdots \subset R_n^* = R^*.$$

But $F^* = F(\omega)$ is a pure extension, so that the radical tower can be lengthened by adding the prefix $F \subset F^*$, thereby displaying R^*/F as a radical extension. •

This theorem implies the classical theorems.

Corollary 99. *If F is a field of characteristic 0, then every polynomial in F[x] of degree n ≤ 4 is solvable by radicals.*

Proof. The Galois group is a subgroup of S_4. But S_4 is solvable, by Theorem G.34, and every subgroup of a solvable group is itself solvable. •

An earlier theorem of Abel states, when translated into group theoretic language, that a polynomial with a commutative Galois group is solvable by radicals; such groups are called *abelian* because of this theorem. Abel's theorem is a special case of Galois's, for every abelian group is solvable.

A deep theorem of Feit and Thompson (1963) says that every group of odd order is solvable. It follows that if F is a field of characteristic 0 and $f(x) \in F(x)$ is a polynomial whose Galois group has odd order, equivalently, whose splitting field has odd degree over F, then $f(x)$ is solvable by radicals.

Suppose one knows the Galois group G of a polynomial $f(x) \in \mathbb{Q}[x]$ and that G is solvable. Can one, in practice, use this information to find the roots of $f(x)$? The answer is affirmative; we suggest the reader look at the books of [Edwards] and [Gaal] to see how this is done.

Exercises

95. Let E/F be a finite separable extension with Galois group G. Define the *trace* $T : E \to E$ by $T(\alpha) = \sum_{\sigma \in G} \sigma(\alpha)$.

    (i) Prove that im $T \subset F$ and that

$$T(\alpha + \beta) = T(\alpha) + T(\beta)$$

    for all $\alpha, \beta \in E$.

    (ii) Show that T is not identically zero. (Hint: Independence of characters.)

96. Assume that E/F is a separable extension of degree n and cyclic Galois group $G = \text{Gal}(E/F) = \langle \sigma \rangle$.

    (i) If $\sigma \in G$, define $\tau = \sigma - $ identity, and prove that ker $T = $ im τ. (Hint: Use E/F being a Galois extension to show that ker $\tau = F$ and hence dim(im τ) $= n - 1$; show that dim(ker τ) $= n - 1$ as well.)

(ii) Prove the **Trace Theorem**: If E/F is a Galois extension with cyclic Galois group $\text{Gal}(E/F) = \langle \sigma \rangle$, then

$$\ker T = \{\alpha \in E : \alpha = \sigma(\beta) - \beta \text{ for some } \beta \in E\}.$$

97. Let F be a field of characteristic $p > 0$.

(i) Let $f(x) = x^p - x - c \in F[x]$ and let u be a root of $f(x)$ in some splitting field E/F. Show that every root of $f(x)$ has the form $u + i$, where $0 \le i < p$.

(ii) Show that $x^p - x - c \in F[x]$ either splits or is irreducible.

98. Let F be a field of characteristic $p > 0$, and let E/F be a Galois extension with cyclic Galois group $\langle \sigma \rangle$ of order p.

(i) Prove there is $\alpha \in E$ with $\sigma(\alpha) - \alpha = 1$. (Hint. Use the trace theorem.)

(ii) Prove that $E = F(\alpha)$, where α is a root of an irreducible polynomial in $F[x]$ of the form $x^p - x - c$.

99. Here is a proof of Exercise 98, similar to that in Corollary 97, which does not use the trace theorem. Let E/F be a Galois extension, where F is a field of characteristic $p > 0$, and let $\sigma \in \text{Gal}(E/F)$ have order p. View σ as a linear transformation, and define $\tau = \sigma - \text{identity}$.

(i) Prove that $\tau^p = 0$.

(ii) Prove that if $\alpha \in \ker \tau + \text{im} \, \tau$, then $\tau^{p-1}(\alpha) = 0$. Using the fact that p and $p - 1$ are relatively prime, prove that $\tau(\alpha) = 0$.

(iii) Prove that $\ker \tau = F$ and that $\text{im} \, \tau \cap \ker \tau \neq \{0\}$. (Hint. Show that $E = \text{im} \, \tau + \ker \tau$ if $\text{im} \, \tau \cap \ker \tau = \{0\}$.)

(iv) Prove that $1 \in \text{im} \, \tau$. (Hint. Prove that $\text{im} \, \tau \cap \ker \tau = F$, and so $F \subset \text{im} \, \tau$.)

Discriminants

Let F be a field of characteristic 0, let $f(x) \in F[x]$ be a polynomial of degree n having splitting field E/F, and let $G = \text{Gal}(E/F)$. If

$$f(x) = c(x - \alpha_1) \cdots (x - \alpha_n),$$

define

$$\Delta = \prod_{i<j}(\alpha_i - \alpha_j).$$

Although the number Δ does depend on the indexing of the roots, a new indexing of the roots can only change some factors $\alpha_i - \alpha_j$ to $\alpha_j - \alpha_i = -(\alpha_i - \alpha_j)$. Therefore, the sign of Δ depends on the listing of the roots, whereas $D = \Delta^2$ depends only on the set of roots.

Remark. There is a connection between Δ and the alternating group A_n. If $\pi \in S_n$, let π act on $\Delta = \prod_{i<j}(\alpha_i - \alpha_j)$ by permuting the subscripts: $\pi(\Delta) = \prod_{i<j}(\alpha_{\pi i} - \alpha_{\pi j})$. Now $\pi(\Delta) = \pm\Delta$; define $\theta : S_n \to \mathbb{Z}_2$ by $\theta(\pi) = [0]$ if $\pi(\Delta) = \Delta$, and $\theta(\pi) = [1]$ if $\pi(\Delta) = -\Delta$. It is easy to see that θ is a surjective homomorphism with kernel A_n, for the alternating group is the unique subgroup of S_n having index 2 (Theorem G.29). Therefore, $\pi(\Delta) = \Delta$ for π even, and $\pi(\Delta) = -\Delta$ for π odd.

Definition. The *discriminant* of a polynomial $f(x) \in F[x]$ is $D = \Delta^2$.

It is clear that $f(x)$ has repeated roots if and only if $D = 0$. Each $\sigma \in G$ permutes $\alpha_1, \ldots, \alpha_n$, so that $\sigma(\Delta) = \pm\Delta$; hence $\Delta^2 = D \in E^G = F$.

If $f(x) = x^2 + bx + c$, then the quadratic formula gives the roots of $f(x)$:

$$\alpha = \tfrac{1}{2}\left(-b + \sqrt{b^2 - 4c}\right) \quad \text{and} \quad \beta = \tfrac{1}{2}\left(-b - \sqrt{b^2 - 4c}\right).$$

It follows that

$$D = \Delta^2 = (\alpha - \beta)^2 = b^2 - 4c.$$

If $f(x)$ is a cubic with roots α, β, γ, then

$$D = \Delta^2 = (\alpha - \beta)^2(\alpha - \gamma)^2(\beta - \gamma)^2;$$

it is not obvious how to compute D from the coefficients of $f(x)$.

Definition. A polynomial $f(x) = x^n + c_{n-1}x^{n-1} + \cdots + c_0$ is *reduced* if $c_{n-1} = 0$. If $f(X)$ is a monic polynomial of degree n and if $c_n \neq 0$ in F, then its *associated reduced polynomial* is

$$\tilde{f}(x) = f(x - c_{n-1}/n).$$

Recall Lemma 43: If $f(x) = x^n + c_{n-1}x^{n-1} + \cdots + c_0 \in F[x]$ and $\beta \in F$, then β is a root of $\tilde{f}(x)$ if and only if $\beta - c_{n-1}/n$ is a root of $f(x)$.

Theorem 100. (i) *A polynomial $f(x)$ and its associated reduced polynomial $\tilde{f}(x)$ have the same discriminant.*

(ii) *The discriminant of a reduced cubic $\tilde{f}(x) = x^3 + qx + r$ is*

$$D = -4q^3 - 27r^2.$$

Proof. (i) If the roots of $f(x)$ are $\alpha_1, \ldots, \alpha_n$, then the roots of $\tilde{f}(x)$ are $\beta_1, \ldots, \beta_n$, where $\beta_i = \alpha_i - a_{n-1}/n$. Therefore

$$\prod_{i<j}(\alpha_i - \alpha_j) = \prod_{i<j}(\beta_i - \beta_j),$$

and so the discriminants (which are the squares of these) are equal.

(ii) The cubic formula gives the following roots of $\tilde{f}(x)$:

$$\alpha = y + z; \quad \beta = \omega y + \omega^2 z; \quad \gamma = \omega^2 y + \omega z;$$

here, $y = \left[\frac{1}{2}(-r + \sqrt{R})\right]^{1/3}$, $z = -q/3y$, ω is a cube root of unity, and $R = r^2 + 4q^3/27$. Because $\omega^3 = 1$, we have

$$\begin{aligned}
\alpha - \beta &= y + z - \omega y - \omega^2 z \\
&= (y - \omega^2 z) - (\omega y - z) \\
&= (y - \omega^2 z) - (y - \omega^2 z)\omega \\
&= (y - \omega^2 z)(1 - \omega).
\end{aligned}$$

Similar calculations give

$$\alpha - \gamma = y + z - \omega^2 y - \omega z = (y - \omega z)(1 - \omega^2)$$

and

$$\beta - \gamma = \omega y + \omega^2 z - \omega^2 y - \omega z = (y - z)\omega(1 - \omega).$$

It follows that

$$\Delta = (y - z)(y - \omega z)(y - \omega^2 z)\omega(1 - \omega^2)(1 - \omega)^2.$$

By Exercise 100, we have $\omega(1 - \omega^2)(1 - \omega)^2 = 3i\sqrt{3}$ (where $i^2 = -1$); moreover, the identity

$$x^3 - 1 = (x - 1)(x - \omega)(x - \omega^2),$$

with $x = y/z$, gives

$$(y - z)(y - \omega z)(y - \omega^2 z) = y^3 - z^3 = \sqrt{R}.$$

Therefore, $\Delta = 3i\sqrt{3}\sqrt{R}$ and

$$D = \Delta^2 = -27R = -27r^2 - 4q^3. \quad \bullet$$

Corollary 101. *Let $f(x) = x^3 + qx + r \in \mathbb{C}[x]$ have discriminant D and roots u, v and w. If $F = \mathbb{Q}(q, r)$, then $F(u, \sqrt{D})$ is a splitting field of $f(x)$ over F.*

Proof. Let $E = F(u, v, w)$ be a splitting field of $f(x)$, and let $K = F(u, \sqrt{D})$. Now $K \subset E$, for the definition of discriminant gives $\sqrt{D} = \pm(u-v)(u-w)(v-w) \in E$. For the reverse inclusion, it suffices to prove that $v \in K$ and $w \in K$.

Since $u \in K$ is a root of $f(x)$, there is a factorization

$$f(x) = (x - u)g(x) \text{ in } K[x].$$

Now the roots of the quadratic $g(x)$ are v and w, so that

$$g(x) = (x - v)(x - w) = x^2 - (v + w)x + vw.$$

Since $g(x)$ has its coefficients in K and since $u \in K$, we have

$$g(u) = (u - v)(u - w) \in K.$$

Therefore,

$$\begin{aligned} v - w &= (u - v)(u - w)(v - w)/(u - v)(u - w) \\ &= \pm\sqrt{D}/(u - v)(u - w) \in K. \end{aligned}$$

On the other hand, $v + w \in K$ because it is a coefficient of $g(x)$, so that

$$v + w \in K \text{ and } v - w \in K.$$

It follows that $v, w \in K$, and so $E = F(u, v, w) \subset K = F(u, \sqrt{D})$.   ●

In Example 17, we observed that the cubic formula giving the roots of $f(x) = x^3 + qx + r$ involves $\sqrt{R}$, where $R = r^2 + 4q^3/27$. Thus, when R is negative, every root of $f(x)$ involves complex numbers. Since the discriminant $D = -27R$, real roots are given in terms of complex numbers whenever $D > 0$. This phenomenon was quite disquieting to mathematicians of the sixteenth century, who spent much time trying to rewrite specific formulas to eliminate complex numbers. The next theorem shows why they were doomed to fail in their rewriting attempts. Even though these attempts were unsuccessful, they ultimately led to a greater understanding of complex numbers.

Theorem 102 (**Casus Irreducibilis**). *Let $f(x) = x^3 + qx + r \in \mathbb{Q}[x]$ be an irreducible cubic having real roots u, v, and w. Let $F = \mathbb{Q}(q, r)$, let $E = F(u, v, w)$ be a splitting field of $f(x)$, and let*

$$F = K_0 \subset K_1 \subset \cdots \subset K_t$$

be a radical tower with $E \subset K_t$. Then K_t is never a subfield of $\mathbb{R}$.

Proof. Since all the roots u, v and w are real,

$$D = [(u - v)(u - w)(v - w)]^2 \geq 0,$$

and so $\sqrt{D}$ is real. There is no loss in generality in assuming that $\sqrt{D}$ has been adjoined first:

$$K_1 = F(\sqrt{D}).$$

We claim that $f(x)$ remains irreducible in $K_1[x]$. If not, then K_1 contains a root of $f(x)$, say, u. If v and w are the other roots of $f(x)$, then $w \in K_1(v)$ because $x - w = f(x)/(x - u)(x - v) \in K_1(v)[x]$, and hence $E \subset K_1(v)$. The reverse inclusion holds, for E contains v and $\sqrt{D}$; thus, $E = K_1(v)$. Now $[E : K_1] \leq 2$ and $[K_1 : F] \leq 2$, so that $[E : F] = [E : K_1][K_1 : F]$ is a divisor of 4. By Exercise 78(i), the irreducibility of $f(x)$ over F gives $3 \mid [E : F]$. This contradiction shows that $f(x)$ is irreducible in $K_1[x]$.

We may assume, by Exercise 82, that each pure extension K_{i+1}/K_i in the radical tower is of prime type. As $f(x)$ is irreducible in $K_1[x]$ and splits in $K_t[x]$ (because $E \subset K_t$), there is a first pure extension K_{j+1}/K_j with $f(x)$ irreducible over K_j and factoring over K_{j+1}. By hypothesis, $K_{j+1} = K_j(\alpha)$, where α is a root of $x^p - c$ for some prime p and some $c \in K_j$. By Corollary 72, either $x^p - c$ is irreducible over K_j or c is a pth power in K_j. In the latter case, we have $K_{j+1} = K_j$, contradicting $f(x)$ being irreducible over K_j but not over K_{j+1}. Therefore, $x^p - c$ is irreducible over K_j, so that

$$[K_{j+1} : K_j] = p,$$

by Theorem 45. Since $f(x)$ factors over K_{j+1}, there is a root of $f(x)$ lying in it, say,

$$u \in K_{j+1};$$

hence, $K_j \subset K_j(u) \subset K_{j+1}$. But $f(x)$ is an irreducible cubic over K_j, so that $3 \mid [K_{j+1} : K_j] = p$, by Exercise 78(i). It follows that $p = 3$ and

$$K_{j+1} = K_j(u).$$

Now K_{j+1} contains u and $\sqrt{D}$, so that $K_j \subset E = F(u, \sqrt{D}) \subset K_{j+1}$, by Corollary 101. Since $[K_{j+1} : K_j]$ has no proper intermediate subfields (Exercise 78(i) again), we have $K_{j+1} = E$. Thus, K_{j+1} is a splitting field of $f(x)$ over K_j, and hence K_{j+1} is a Galois extension of K_j. The polynomial $x^3 - c$ (remember that $p = 3$) has a root, namely α, in K_{j+1}, so that Theorem 81 says that K_{j+1} contains the other roots $\omega\alpha$ and $\omega^2\alpha$ as well, where ω is a primitive cube root of unity. But this gives $\omega = (\omega\alpha)/\alpha \in K_{j+1}$, which is a contradiction because ω is not real while $K_{j+1} \subset K_t \subset \mathbb{R}$. •

Exercises

100. Prove that $\omega(1 - \omega^2)(1 - \omega)^2 = 3i\sqrt{3}$.

101. (i) Prove that if $a \neq 0$, then $f(x)$ and $af(x)$ have the same discriminant and the same Galois group. Conclude that it is no loss in generality to restrict attention to monic polynomials when computing Galois groups.

(ii) Prove that a polynomial $f(x)$ and its associated reduced polynomial $\tilde{f}(x)$ have the same Galois group.

102. (i) If $f(x) = x^3 + ax^2 + bx + c$, then its associated reduced polynomial is $x^3 + qx + r$, where

$$q = b - a^2/3 \quad \text{and} \quad r = 2a^3/27 - ab/3 + c.$$

(ii) Show that the discriminant of $f(x)$ is

$$D = a^2b^2 - 4b^3 - 4a^3c - 27c^2 + 18abc.$$

Galois Groups of Quadratics, Cubics, and Quartics

In this final section, we show how to compute Galois groups of polynomials of low degree over $\mathbb{Q}$. Recall that the Galois group of a polynomial of degree n is a subgroup of S_n (regarded as the group of all permutations of the roots). Of course, just as there are some permutations of the vertices of a polygon that do not arise from symmetries, so, too, some permutations of the roots of a polynomial may have nothing to do with field automorphisms.

Lemma 103. *Let $f(x) \in F[x]$ have discriminant $D = \Delta^2$ and Galois group $G = \mathrm{Gal}(E/F)$. If $H = G \cap A_n$, then $E^H = F(\Delta)$; moreover, $\sqrt{D} \in F$ if and only if G is a subgroup of A_n.*

Proof. Clearly $F(\Delta) \subset E^H$ and $[E^H : F] = [G : H] \leq 2$; it suffices to prove that $[F(\Delta) : F] = [G : H]$. If $[G : H] = 2$, then there exists $\sigma \in G, \sigma \notin H$, with $\sigma(\Delta) \neq \Delta$; hence $\Delta \notin E^G = F$ and $[F(\Delta) : F] = 2$. If $[G : H] = 1$, then $G = H$ and $F(\Delta) \subset E^H = E^G = F$; hence $[F(\Delta) : F] = 1$.

For the second statement, the Fundamental Theorem of Galois Theory says that $F(\Delta) = E^H = F$ if and only if $G = H$ (because $E^G = F$). Since $H = G \cap A_n$, this means that $G \subset A_n$. •

If $f(x) \in \mathbb{Q}[x]$ is quadratic, then its Galois group has order either 1 or 2 (because the symmetric group S_2 has order 2). The Galois group has order 1 if $f(x)$ splits; it has order 2 if $f(x)$ does not split; that is, if $f(x)$ is irreducible.

If $f(x) \in \mathbb{Q}[x]$ is a cubic having a rational root, then its Galois group G is the same as that of its quadratic factor. Otherwise $f(x)$ is irreducible; since $|G|$ is now a multiple of 3, by Exercise 78(ii), and $G \subset S_3$, it follows that either $G \cong A_3 \cong \mathbb{Z}_3$ or $G \cong S_3$.

Theorem 104. *Let $f(x) \in \mathbb{Q}[x]$ be an irreducible cubic with Galois group G and discriminant D.*

   (i) $f(x)$ *has exactly one real root if and only if $D < 0$, in which case $G \cong S_3$.*

   (ii) $f(x)$ *has three real roots if and only if $D > 0$. In this case, either $\sqrt{D} \in \mathbb{Q}$ and $G \cong \mathbb{Z}_3$ or $\sqrt{D} \notin \mathbb{Q}$ and $G \cong S_3$.*

Proof. Note first that $D \neq 0$: since $\mathbb{Q}$ has characteristic 0, it is perfect, and hence irreducible polynomials over $\mathbb{Q}$ have no repeated roots. If $f(x)$ has three real roots, then Δ is real and $D = \Delta^2 > 0$. Conversely assume that $f(x)$ has one real root α and two complex roots: $\beta = u+iv$ and $\bar{\beta} = u-iv$. Since $\beta - \bar{\beta} = 2iv$ and $\alpha = \bar{\alpha}$, we have

$$
\begin{aligned}
\Delta &= (\alpha - \beta)(\alpha - \bar{\beta})(\beta - \bar{\beta}) \\
&= (\alpha - \beta)\overline{(\alpha - \beta)}(\beta - \bar{\beta}) \\
&= |\alpha - \beta|^2(2iv),
\end{aligned}
$$

and so $D = \Delta^2 = -4v^2|\alpha - \beta|^4 < 0$.

Let $E/\mathbb{Q}$ be the splitting field of $f(x)$. If $f(x)$ has exactly one real root α, then $E \neq \mathbb{Q}(\alpha)$. Hence $|G| > 3$ and $G \cong S_3$. If $f(x)$ has three real roots, then $D > 0$ and $\sqrt{D}$ is real. By Lemma 103, $G \cong A_3 \cong \mathbb{Z}_3$ if and only if $\sqrt{D}$ is rational; hence $G \cong S_3$ if $\sqrt{D}$ is irrational.   •

Example 35. The polynomial $x^3 - 2 \in \mathbb{Q}[x]$ is irreducible, and its discriminant is $D = -108$. Therefore, its Galois group is S_3.

The polynomial $x^3 - 4x + 2 \in \mathbb{Q}[x]$ is irreducible, by Eisenstein's criterion, and its discriminant is $D = 148$. Since $\sqrt{148}$ is irrational, the Galois group is S_3.

The polynomial $x^3 - x + \frac{1}{3} \in \mathbb{Q}[x]$ is irreducible, by Exercise 63, and its discriminant is $D = 1$. Since $\sqrt{1}$ is rational, the Galois group is $\mathbb{Z}_3$.

Consider a (reduced) quartic $f(x) = x^4 + qx^2 + rx + s \in \mathbb{Q}[x]$; let $E/\mathbb{Q}$ be its splitting field and let $G = \text{Gal}(E/\mathbb{Q})$ be its Galois group. (By Exercise 101, it is no loss in generality to assume $f(x)$ is reduced.) If $f(x)$ has a rational root u, then $f(x) = (x - u)c(x)$, and its Galois group is the same as that of its cubic factor $c(x)$; but Galois groups of cubics have already been discussed. Suppose that $f(x) = h(x)k(x)$ is the product of two irreducible quadratics; let α be a root of $h(x)$ and let β be a root of $k(x)$. If $\mathbb{Q}(\alpha) \cap \mathbb{Q}(\beta) = \mathbb{Q}$, that is, if these fields are linearly disjoint, then Corollary 91 shows that $G \cong V$, the four group; otherwise, $\alpha \in \mathbb{Q}(\beta)$, so that $\mathbb{Q}(\beta) = \mathbb{Q}(\alpha, \beta) = E$, and G has order 2.

We are left with the case $f(x)$ irreducible. The basic idea now is to compare G with the four group, namely, the normal subgroup of S_4:

$$V = \{(1), (12)(34), (13)(24), (14)(23)\},$$

so that we can identify the fixed field of $V \cap G$. If the four (distinct) roots of $f(x)$ are $\alpha_1, \alpha_2, \alpha_3, \alpha_4$, then consider the numbers:

$$\begin{aligned} u &= (\alpha_1 + \alpha_2)(\alpha_3 + \alpha_4), \\ v &= (\alpha_1 + \alpha_3)(\alpha_2 + \alpha_4), \\ w &= (\alpha_1 + \alpha_4)(\alpha_2 + \alpha_3). \end{aligned}$$

It is clear that if $\sigma \in V \cap G$, then σ fixes u, v, and w. Conversely, checking each of the 24 permutations shows that if $\sigma \in S_4$ fixes $(\alpha_i + \alpha_j)(\alpha_k + \alpha_\ell)$, then $\sigma \in V \cup \{(ij), (k\ell), (ikj\ell), (i\ell jk)\}$. It follows that $\sigma \in G$ fixes each of u, v, w if and only if $\sigma \in V \cap G$, and so $\mathbb{Q}(u, v, w)$ is the fixed field of $V \cap G$.

Definition. The *resolvent cubic* [17] of $f(x) = x^4 + qx^2 + rx + s$ is

$$g(x) = (x - u)(x - v)(x - w).$$

Theorem 105. *If $g(x)$ is the resolvent cubic of $f(x) = x^4 + qx^2 + rx + s$,* then

$$g(x) = x^3 - 2qx^2 + (q^2 - 4s)x + r^2.$$

Proof. In our discussion of the classical quartic formula, we saw that $f(x) = (x^2 + kx + \ell)(x^2 - kx + m)$ and k^2 is a root of

$$h(x) = x^3 + 2qx^2 + (q^2 - 4s)x - r^2,$$

a polynomial differing from the claimed expression for $g(x)$ only in the sign of its quadratic and constant terms. Thus, a number β is a root of $h(x)$ if and only if $-\beta$ is a root of $g(x)$.

Let the four roots $\alpha_1, \alpha_2, \alpha_3, \alpha_4$ of $f(x)$ be indexed so that α_1, α_2 are roots of $x^2 + kx + \ell$ and α_3, α_4 are roots of $x^2 - kx + m$. Then $k = -(\alpha_1 + \alpha_2)$ and $-k = -(\alpha_3 + \alpha_4)$; therefore

$$u = (\alpha_1 + \alpha_2)(\alpha_3 + \alpha_4) = -k^2$$

and $-u$ is a root of $h(x)$ since $h(k^2) = 0$.

Now factor $f(x)$ into two quadratics, say,

$$f(x) = (x^2 + \tilde{k}x + \tilde{\ell})(x^2 - \tilde{k}x + \tilde{m}),$$

where α_1, α_3 are roots of the first factor and α_2, α_4 are roots of the second. The same argument as above now shows that

$$v = (\alpha_1 + \alpha_3)(\alpha_2 + \alpha_3) = -\tilde{k}^2,$$

---

[17]There is another resolvent cubic in the literature which arises from another combination of the roots invariant under V. Define

$$
\begin{aligned}
u' &= \alpha_1\alpha_2 + \alpha_3\alpha_4, \\
v' &= \alpha_1\alpha_3 + \alpha_2\alpha_4, \\
w' &= \alpha_1\alpha_4 + \alpha_2\alpha_3,
\end{aligned}
$$

and define $h(x) = (x - u')(x - v')(x - w')$. This cubic (which is distinct from $g(x)$ above) behaves much as $g(x)$ does. The reason for our preference for $g(x)$ is Exercise 103; one can use $g(x)$ to compute the discriminant of a quartic.

hence $-v$ is a root of $h(x)$. Similarly, $-w = -(\alpha_1 + \alpha_4)(\alpha_2 + \alpha_3)$ is a root of $h(x)$. Therefore

$$h(x) = (x + u)(x + v)(x + w),$$

and so

$$g(x) = (x - u)(x - v)(x - w)$$

is obtained from $h(x)$ by changing the sign of the quadratic and constant terms. •

Theorem 106. *Let $f(x) \in \mathbb{Q}[x]$ be an irreducible quartic with Galois group G, and let m be the order of the Galois group of its resolvent cubic.*

(i) *If $m = 6$, then $G \cong S_4$.*

(ii) *If $m = 3$, then $G \cong A_4$.*

(iii) *If $m = 1$, then $G \cong V$.*

(iv) *If $m = 2$, then $G \cong D_8$ or $G \cong \mathbb{Z}_4$.*

Remark. Note that, in the ambiguous case (iv), the two possible groups have different orders. See Exercise 113.

Proof. We have seen that $\mathbb{Q}(u, v, w)$ is the fixed field of $V \cap G$. By the Fundamental Theorem,

$$
\begin{aligned}
|G/V \cap G| &= [G : V \cap G] \\
&= [\mathbb{Q}(u, v, w) : \mathbb{Q}] \\
&= |\operatorname{Gal}(\mathbb{Q}(u, v, w)/\mathbb{Q})|.
\end{aligned}
$$

Since $f(x)$ is irreducible, G acts transitively on its roots, by Exercise 79, hence $|G|$ is divisible by 4 (Theorem G.10), and the theorem follows from Exercise 106 and Exercise 107. •

Example 36. Let $f(x) = x^4 - 4x + 2 \in \mathbb{Q}[x]$; $f(x)$ is irreducible, by Eisenstein's criterion. The resolvent cubic is

$$g(x) = x^3 - 8x + 16.$$

Now $g(x)$ is irreducible, for if one reduces mod 5, one obtains $x^3 + 2x + 1$, and this polynomial is irreducible over $\mathbb{Z}_5$ because it has no roots. The discriminant of $g(x)$ is -4864, so that Theorem 104 shows that the Galois group of $g(x)$ is S_3, hence has order 6. Theorem 106 now shows that $G \cong S_4$.

Example 37. Let $f(x) = x^4 - 10x^2 + 1 \in \mathbb{Q}[x]$; $f(x)$ is irreducible, by Exercise 67. The resolvent cubic is

$$x^3 + 20x^2 + 96x = x(x + 8)(x + 12).$$

In this case, $\mathbb{Q}(u, v, w) = \mathbb{Q}$ and $m = 1$. Therefore, $G \cong V$. (This should not be a surprise if one recalls Example 20 where we saw that $f(x)$ arises as the irreducible polynomial of $\alpha = \sqrt{2} + \sqrt{3}$, where $\mathbb{Q}(\alpha) = \mathbb{Q}(\sqrt{2}, \sqrt{3})$.)

Remark. If d is a divisor of $|S_4| = 24$, then it is known that S_4 has a subgroup of order d. If $d = 4$, then V and $\mathbb{Z}_4$ are nonisomorphic subgroups of order d; for any other divisor d, any two subgroups of order d are isomorphic. We conclude that the Galois group G of a quartic is determined to isomorphism by its order unless $|G| = 4$.

Exercises

103. If $f(x)$ is a quartic, then its discriminant is the discriminant of its resolvent cubic. (Hint:

$$
\begin{aligned}
u - v &= -(\alpha_1 - \alpha_4)(\alpha_2 - \alpha_3) \\
u - w &= -(\alpha_1 - \alpha_3)(\alpha_2 - \alpha_4) \\
v - w &= -(\alpha_1 - \alpha_2)(\alpha_3 - \alpha_4).)
\end{aligned}
$$

104. If $f(x) = x^4 + ax^2 + bx + c$, prove that the discriminant of $f(x)$ is

$$D = -16a^4c + 4a^3b^2 + 128a^2c^2 - 144ab^2c + 27b^4 - 256c^3.$$

105. Show that $x^3 + ax + 2 \in \mathbb{R}[x]$ has three real roots if and only if $a \leq -3$.

106. Let G be a subgroup of S_4 with $|G|$ a multiple of 4; define

$$m = |G/G \cap V|,$$

where $V = \{1, (12)(34), (13)(24), (14)(23)\}$ is the four group.

   (i) Prove that m is a divisor of 6.

   (ii) If $m = 6$, then $G = S_4$; if $m = 3$, then $G = A_4$; if $m = 1$, then $G = V$; if $m = 2$, then $G \cong D_8$, $G \cong \mathbb{Z}_4$, or $G \cong V$. (Hint: This exercise in group theory is Theorem G.35.)

107. Let G be a subgroup of S_4. If G acts transitively on $X = \{1, 2, 3, 4\}$ and $|G/G \cap V| = 2$, then $G \cong D_8$ or $G \cong \mathbb{Z}_4$. (If we merely assume that G acts transitively on X, then $|G|$ is a multiple of 4 (Theorem G.10). The added hypothesis $|G/G \cap V| = 2$ removes the possibility $G \cong V$ when $m = 2$ in Exercise 106.)

108. Compute the Galois group over $\mathbb{Q}$ of $x^4 + x^2 - 6$.

109. Compute the Galois group over $\mathbb{Q}$ of $f(x) = x^4 + x^2 + x + 1$.

110. Compute the Galois group over $\mathbb{Q}$ of $f(x) = 4x^4 + 12x + 9$. (Hint. Prove that $f(x)$ is irreducible in two steps: first show it has no rational roots, and then use Descartes's method for the quartic formula and Exercise 64 to show that $f(x)$ is not the product of two quadratics over $\mathbb{Q}$.)

111.   (i) Prove that a quintic polynomial over $\mathbb{Q}$ is solvable by radicals if and only if its Galois group has order ≤ 24.

   (ii) Prove that an irreducible quintic over $\mathbb{Q}$ is solvable by radicals if and only if its Galois group has order ≤ 20. (Hint: A subgroup G of S_5 is solvable if and only if $|G| \leq 24$; see Theorem G.40.)

112. (Kaplansky) Let $f(x) \in \mathbb{Q}[x]$ be an irreducible quartic with Galois group G. If $f(x)$ has exactly two real roots, then either $G \cong S_4$ or $G \cong D_8$.

113. (Kaplansky) Let $x^4 + ax^2 + b$ be an irreducible polynomial over $\mathbb{Q}$ having Galois group G.

   (i) If b is a square in $\mathbb{Q}$, then $G \cong V$.

   (ii) If b is not a square in $\mathbb{Q}$ but $b(a^2 - 4b)$ is a square, then $G \cong \mathbb{Z}_4$.

   (iii) If neither b nor $b(a^2 - 4b)$ is a square, then $G \cong D_8$.

114. (Kaplansky) Let $x^4 + bx^3 + cx^2 + bx + 1 \in \mathbb{Q}[x]$ have Galois group G.

   (i) If $h = c^2 + 4c + 4 - 4b^2$ is a square in $\mathbb{Q}$, then $G \cong V$.

   (ii) If h is not a square in $\mathbb{Q}$ but $h(b^2 - 4c + 8)$ is a square, then $G \cong \mathbb{Z}_4$.

   (iii) If neither h nor $h(b^2 - 4c + 8)$ is a square in $\mathbb{Q}$, then $G \cong D_8$.

115. If a herring and a half cost a penny and a half, how much does a dozen herring cost? (Answer: One shilling.)

Epilogue

You have seen an introduction to Galois theory; of course, there is more. A deeper study of *abelian fields,* that is, fields having (possibly infinite) abelian Galois groups, begins with *Kummer theory* and continues on to *class field theory.* Infinite Galois groups are topologized, and there is a bijection between intermediate fields and closed subgroups. The theorems are of basic importance in algebraic number theory. There is also a Galois theory classifying division algebras (see [Jacobson (1956)] and a Galois theory classifying commutative rings (see [Chase, Harrison, Rosenberg]).

An interesting open question is to determine which abstract finite groups G can be realized as Galois groups over $\mathbb{Q}$ (Exercise 91 shows that G can always be realized over some ground field). Many special examples have long been known. For example, Hilbert proved that the symmetric groups can be realized over $\mathbb{Q}$ (a proof can be found in [Hadlock, p. 210]); for a proof that the quaternions can be realized, see [R.A. Dean, Amer. Math. Monthly (1981), pp. 42–45] where it is shown to be the Galois group of $x^8 - 72x^6 + 180x^4 - 144x^2 + 36$). It is a deep result of Shafarevich (1954) that every solvable group can be realized. After the classification of the finite simple groups in the 1980's, there were attempts to realize them, with much success. However, it is still not known whether every finite simple group is a Galois group over $\mathbb{Q}$.

There is Galois theory in complex variables (see [Miller, Blichfeldt, and Dickson, Chap. XX, p. 378]). In 1850, Puiseux studied the *monodromy group* of certain functions of two complex variables, namely, $f(t, z) \in \mathbb{C}(t)[z]$; in 1851, Hermite showed that this monodromy group is isomorphic to the Galois group of $f(t, z)$ over the function field $\mathbb{C}(t)$.

There is Galois theory in differential equations, due to Ritt and Kolchin (see [Kaplansky (1957)]). A *derivation* of a field F is an additive homomorphism $D : F \to F$ with $D(xy) = xD(y) + D(x)y$; an ordered pair (F, D) is called a *differential field.* Given a differential field (F, D) with F a (possibly infinite) extension of $\mathbb{C}$, its *differential Galois group* is the subgroup of $\mathrm{Gal}(F/\mathbb{C})$ consisting of all σ commuting with D. If this group is suitably topologized and if the extension $F/\mathbb{C}$ satisfies conditions analogous to being a Galois extension (it is called a *Picard–Vessiot extension*), then there is a bijection between the intermediate differential fields and the closed subgroups of the differential Galois group. The latest developments are in Magid (1994).

There is Galois theory in algebraic topology. A *covering space* of a topological space X is an ordered pair $(\widetilde{X}, p)$, where $p : \widetilde{X} \to X$ is a certain type of continuous map. The elements of the group $\mathrm{Cov}(\widetilde{X}/X)$ defined as {homeomorphisms $h : \widetilde{X} \to \widetilde{X} : ph = p$} are dual to the elements of a Galois group in the following sense. If $i : F \to E$ is the inclusion, where E/F is a Galois extension, then an automorphism σ of E lies in the Galois group if and only if $\sigma i = i$. When $\widetilde{X}$ is simply connected, then $\mathrm{Cov}(\widetilde{X}/X) \cong \pi_1(X)$, the fundamental group of X; moreover, there is a bijection between the family of all covering spaces of X and the family of all subgroups of the fundamental group.

I am awed by the genius of Galois (1811–1832). He solved one of the outstanding mathematical problems of his time, and his solution is beautiful; in so doing, he created two powerful theories, group theory and Galois theory, and his work is still influential today. And he did all of this at the age of 19; he was killed a year later.

Appendices

Appendix A
Group Theory Dictionary

Abelian group. A group in which multiplication is commutative.

Alternating group A_n. The subgroup of S_n consisting of all the even permutations. It has order $\frac{1}{2}n!$.

Associativity. For all x, y, z, one has $(xy)z = x(yz)$. It follows that one does not need parentheses for any product of three or more factors.

Automorphism. An isomorphism of a group with itself.

Commutativity. For all x, y, one has $xy = yx$.

Conjugate elements. Two elements x and y in a group G are called conjugate if there exists $g \in G$ with $y = gxg^{-1}$.

Conjugate subgroups. Two subgroups H and K of a group G are called conjugate if there exists $g \in G$ with

$$K = gHg^{-1} = \{ghg^{-1} : h \in H\}.$$

Coset of H in G. A subset of G of the form $gH = \{gh : h \in H\}$, where H is a subgroup of G and $g \in G$. All the cosets of H partition G; moreover, $gH = g'H$ if and only if $g^{-1}g' \in H$.

Cyclic group. A group G which contains an element g (called a *generator*) such that every element of G is some power of g.

Dihedral group D_{2n}. A group of order $2n$ containing an element a of order n and an element b of order 2 such that $bab = a^{-1}$.

Even permutation. A permutation that is a product of an even number of transpositions. Every r-cycle, for r odd, is an even permutation.

Factor groups. Given a normal series $G = G_0 \supset G_1 \supset \ldots \supset G_n = \{1\}$, its factor groups are the groups G_i/G_{i+1} for $i \geq 0$.

Four group V. The subgroup of S_4 consisting of

$$1, (12)(34), (13)(24), \text{ and } (14)(23);$$

it is a normal subgroup.

Generator of a cyclic group G. An element $g \in G$ whose powers give all the elements of G; a cyclic group may have several different generators.

Group. A set G equipped with an associative multiplication such that there is a unique $e \in G$ (called the *identity* of G) with $ex = x = xe$ for all $x \in G$, and, for each $x \in G$, there is a unique $y \in G$ (called the *inverse* of x) with $yx = e = xy$. One usually denotes e by 1 and y by x^{-1}. (Some of these axioms are redundant.)

Homomorphism. A function $f : G \to H$, where G and H are groups, such that $f(xy) = f(x)f(y)$ for all $x, y \in G$. One always has $f(1) = 1$ and $f(x^{-1}) = f(x)^{-1}$.

Image. Given a homomorphism $f : G \to H$, its image im f is the subgroup of H consisting of all $f(x)$ for $x \in G$.

Index $[G : H]$. The number of (left) cosets of a subgroup H in G; it is equal to $|G|/|H|$ when G is finite.

Isomorphism. A homomorphism that is a bijection.

Kernel. Given a homomorphism $f : G \to H$, its kernel ker f is the (necessarily) normal subgroup of G consisting of all $x \in G$ with $f(x) = 1$. One denotes this by $H \triangleleft G$.

Natural map. If H is a normal subgroup of G, then the natural map is the homomorphism $\pi : G \to G/H$ defined by $\pi(x) = xH$.

Normal series of G. A sequence of subgroups

$$G = G_0 \supset G_1 \supset \ldots \supset G_n = \{1\}$$

with each G_{i+1} a normal subgroup of G_i. (A subgroup G_i may not be a normal subgroup of G.)

Normal subgroup. A subgroup H of a group G such that, for all $g \in G$,

$$gHg^{-1} = \{ghg^{-1} : h \in H\} = H.$$

Order of an element $x \in G$. The least positive integer m, if any, such that $x^m = 1$; otherwise infinity.

Order $|G|$ of a group G. The number of elements in G.

p-group. A group in which every element has order some power of the prime p. If G is finite, the $|G|$ is a power of p.

Permutation. A bijection of a set to itself; all the permutations of a set X form a group under composition, denoted by S_X.

Quotient group G/H. If H is a normal subgroup of G, then G/H is the family of all cosets gH of H with multiplication defined by

$$gHg'H = gg'H;$$

the order of G/H is $[G : H]$; the identity element is $1H = H$; the inverse of gH is $g^{-1}H$.

Simple group G. A group $G \neq \{1\}$ whose only normal subgroups are $\{1\}$ and G.

Solvable group. A group having a normal series with abelian factor groups.

Subgroup H of G. A subset of G containing 1 which is closed under multiplication and inverse.

Subgroup generated by a subset X. The smallest subgroup of G containing X, denoted by $\langle X \rangle$, consists of all the products $x_1{}^a x_2{}^b \ldots x_n{}^z$, where $x_i \in X$ and the exponents $a, b, \ldots, z = \pm 1$.

Sylow p-subgroup of a finite group G. A subgroup of G of order p^n, where p^n is the highest power of p dividing $|G|$. Such subgroups always exist, and any two such are conjugate, hence isomorphic.

Symmetric group S_n. The group of all permutations of $\{1, 2, \ldots, n\}$ under composition; it has order $n!$.

Appendix B
Group Theory Used in the Text

All groups in this appendix are assumed to be finite even though several of the theorems hold (perhaps with different proofs) in the infinite case as well. Definitions of terms not defined in this appendix can be found in the dictionary, Appendix A.

Theorem G.1. *Every subgroup S of a cyclic group $G = \langle a \rangle$ is itself cyclic.*

Proof. If $S = \{1\}$, then S is cyclic with generator 1. Otherwise, let m be the least positive integer for which $a^m \in S$; we claim $S = \langle a^m \rangle$. Clearly $\langle a^m \rangle \subset S$. For the reverse inclusion, take $s = a^k \in S$. By the division algorithm, there are integers q and r with $0 \le r < m$ and

$$k = qm + r.$$

But $a^k = a^{qm+r} = (a^m)^q a^r$ gives $a^r \in S$. If $r > 0$, the minimality of m is contradicted; therefore $r = 0$ and $a^k = (a^m)^q \in \langle a^m \rangle$. •

Theorem G.2. (i) *If $a \in G$ is an element of order n, then $a^m = 1$ if and only if $n \mid m$.*
    (ii) *If $G = \langle a \rangle$ is a cyclic group of order n, then a^k is a generator of G if and only if $(k, n) = 1$.*
    (iii) *If $a \in G$ has order n, then the order of a is $|\langle a \rangle|$.*

Proof. (i) Assume that $a^m = 1$. The division algorithm provides integers q and r with $m = nq + r$, when $0 \le r < n$. It follows that $a^r = a^{m-nq} = a^m a^{-nq} = 1$. If $r > 0$, then we contradict n being the smallest positive integer with $a^n = 1$. Hence $r = 0$ and $n \mid m$. Conversely, if $m = nk$, then $a^m = a^{nk} = (a^n)^k = 1^k = 1$.

(ii) Recall that two integers are *relatively prime* if and only if some integral linear combination of them is 1.

If a^k generates G, then $a \in \langle a^k \rangle$, so that $a = a^{kt}$ for some $t \in \mathbb{Z}$. Therefore $a^{kt-1} = 1$; by (i), $n \mid kt - 1$, so there is $v \in \mathbb{Z}$ with $nv = kt - 1$. Therefore, 1 is a linear combination of k and m, and so $(k, n) = 1$.

Conversely, if $(k, n) = 1$, then $nt + ku = 1$ for $t, u \in \mathbb{Z}$; hence

$$a = a^{nt+ku} = a^{nt} a^{ku} = a^{ku} \in \langle a^k \rangle.$$

Therefore every power of a also lies in $\langle a^k \rangle$ and $G = \langle a^k \rangle$.

(iii) The list $1, a, a^2, \ldots, a^{n-1}$ has no repetitions: if there are $i < j$ with $a^i = a^j$, then $a^{j-i} = 1$, contradicting n being the smallest exponent for which $a^n = 1$. Now $\{1, a, a^2, \ldots, a^{n-1}\} \subset \langle a \rangle$, and we let the reader prove the reverse inclusion. It follows that $|\langle a \rangle| = |\{1, a, a^2, \ldots, a^{n-1}\}| = n$. •

Theorem G.3 (Lagrange). *If H is a subgroup of a group G, then*

$$|G| = [G : H]|H|.$$

Proof. The relation on G, defined by $x \sim y$ if $y = xh$ for some $h \in H$, is an equivalence relation whose equivalence classes are the cosets of H. Therefore, the cosets of H in G partition G. Moreover $|H| = |xH|$ for every $x \in G$ (because $h \mapsto xh$ is a bijection), so that $|G|$ is the number of cosets times their common size. •

It follows that $[G : H] = |G|/|H|$. In particular, if H is a normal subgroup of a group G (so that the quotient group G/H is defined), then

$$|G/H| = [G : H] = |G|/|H|$$

when G is finite.

Another consequence of Lagrange's theorem is that the order of $a \in G$ is a divisor of $|G|$, for Theorem G.2 shows that the order of a is the order of the subgroup $\langle a \rangle$. Hence, $a^{|G|} = 1$ for all $a \in G$.

If $f : G \to H$ is a homomorphism, denote the image of f by im f and the kernel of f by ker f.

Lemma G.4. *Let $f : G \to H$ be a homomorphism. Then f is an injection if and only if $\ker f = \{1\}$.*

Proof. If f is an injection, then $x \neq 1$ implies $f(x) \neq f(1) = 1$, and so $x \notin \ker f$. Conversely, assume $\ker f = \{1\}$ and that $f(x) = f(y)$ for $x, y \in G$. Then

$$1 = f(x)f(y)^{-1} = f(x)f(y^{-1}) = f(xy^{-1})$$

and $xy^{-1} \in \ker f = \{1\}$. Hence $x = y$ and f is an injection.   •

Theorem G.5 (First Isomorphism Theorem). *If $f : G \to H$ is a homomorphism, then $\ker f$ is a normal subgroup of G and*

$$G/\ker f \cong \operatorname{im} f.$$

Proof. Let $K = \ker f$. Let us show K is a subgroup. It does contain 1 (because $f(1) = 1$); if $x, y \in K$ (so that $f(x) = 1 = f(y)$), then $f(xy) = f(x)f(y) = 1$ and $xy \in K$; if $x \in K$, then $f(x^{-1}) = f(x)^{-1} = 1$ and $x^{-1} \in K$. Furthermore, the subgroup K is normal: if $x \in K$ and $g \in G$, then $f(gxg^{-1}) = f(g)f(x)f(g)^{-1} = f(g)f(g)^{-1} = 1$ and so $gxg^{-1} \in K$.

Define $\varphi : G/K \to \operatorname{im} f$ by $\varphi(xK) = f(x)$. Now φ is well defined: if $x'K = xK$, then $x' = xk$ for some $k \in K$, and $f(x') = f(xk) = f(x)f(k) = f(x)$. It is routine to check that φ is a homomorphism (because f is) with $\operatorname{im} \varphi = \operatorname{im} f$. Finally, φ is an injection, by Lemma G.4, because $\varphi(xK) = 1$ implies $f(x) = 1$, hence $x \in K$ and $xK = K$.   •

If K, H are subgroups of G, then $K \vee H$ is the smallest subgroup of G containing K and H; that is, $K \vee H$ is the subgroup of G generated by $K \cup H$.

Lemma G.6. *If K and H are subgroups of G with K normal in G, then $K \vee H = KH = \{kh : k \in K \text{ and } h \in H\} = HK$.*

Proof. Clearly $KH \subset K \vee H$. For the reverse inclusion, it suffices to prove that KH is a subgroup, for it does contain $K \cup H$.

Now $khk_1h_1 = k(hk_1h^{-1})hh_1 = (kk_2)(hh_1) \in KH$ for some $k_2 \in K$ (because K is normal). Also $(kh)^{-1} = h^{-1}k^{-1} = (h^{-1}k^{-1}h)h^{-1} = k'h^{-1} \in KH$ for some $k' \in K$ (again, because K is normal). Therefore, KH is a subgroup.

If $hk \in HK$, then $hk = (hkh^{-1})h = k'h \in KH$ for some $k' \in K$, and so $HK \subset KH$; the reverse inclusion is proved similarly.   •

If K and H are subgroups of G with K normal, then the family of those cosets hK of K with $h \in H$ is easily seen to be a subgroup of G/K. Indeed, one may check, using Lemma G.6, that this subgroup is KH/K.

Theorem G.7 (Second Isomorphism Theorem). *If K and H are subgroups of G with K normal in G, then K ∩ H is a normal subgroup of H and*

$$H/(K \cap H) \cong KH/K.$$

Proof. Let $\pi : G \to G/K$ be the natural map, defined by $\pi(x) = xK$, and let $f : H \to G/K$ be the restriction $\pi \mid H$. Now $\ker f = K \cap H$ and $\operatorname{im} f$ is the family of all cosets xK in G/K with $x \in H$ (hence $\operatorname{im} f = KH/K$). The first isomorphism theorem now gives the result.   •

Theorem G.8 (Third Isomorphism Theorem). *If $S \subset K$ are normal subgroups of G, then K/S is a normal subgroup of G/S and*

$$(G/S)/(K/S) \cong G/K.$$

Proof. The function $f : G/S \to G/K$ given by $xS \mapsto xK$ is well defined because $S \subset K$. One checks easily that f is a surjective homomorphism with kernel K/S, and so the theorem follows from the first isomorphism theorem.   •

Theorem G.9 (Correspondence Theorem). *Let K be a normal subgroup of G, and let S^* be a subgroup of $G^* = G/K$.*

(i) *There is a unique intermediate subgroup S, i.e., $K \subset S \subset G$, with $S/K = S^*$;*

(ii) *If S^* is a normal subgroup of G^*, then S is normal in G;*

(iii) $[G^* : S^*] = [G : S]$;

(iv) *If T^* is normal in S^*, then T is normal in S and*

$$S^*/T^* \cong S/T.$$

Proof. (i) Define $S = \{x \in G : xK \in S^*\}$.

(ii) If $a \in G$, and $x \in S$, then $axa^{-1}K = aKxKa^{-1}K \in S^*$, because S^* is normal in G^*; therefore $axa^{-1} \in S$.

(iii)

$$\begin{aligned} [G^* : S^*] &= |G^*|/|S^*| = |G/K|/|S/K| \\ &= (|G|/|K|)/(|S|/|K|) = |G|/|S| = [G : S]. \end{aligned}$$

(iv) T is normal in S, by (ii), and

$$S^*/T^* = (S/K)/(T/K) \cong S/T,$$

by the third isomorphism theorem.   •

Definition. A group G *acts* on a set X if there is a function

$$G \times X \to X,$$

denoted by $(g, x) \mapsto g \cdot x$, such that:

(i) $1 \cdot x = x$ for all $x \in X$, where 1 is the identity in G;

(ii) $(gh) \cdot x = g \cdot (h \cdot x)$ for all $x \in X$ and for all $g, h \in G$.

Definition. If G acts on X and $x \in X$, then the *orbit* of x is

$$\mathcal{O}(x) = \{g \cdot x : g \in G\} \subset X,$$

and the *stabilizer* of x is the subgroup

$$G_x = \{g \in G : g \cdot x = x\} \subset G.$$

A group G acts *transitively* on X if, for each $x, y \in X$, there exists $g \in G$ with $g \cdot x = y$. In this case, $\mathcal{O}(x) = X$.

Every group G acts on itself (here $X = G$) by conjugation: define

$$g \cdot x = gxg^{-1}.$$

The orbit $\mathcal{O}(x)$ of $x \in G$ is its *conjugacy class*:

$$\{y \in G : y = gxg^{-1} \quad \text{for some } g \in G\};$$

the stabilizer of x is

$$\{g \in G : x = g \cdot x = gxg^{-1}\} = \{g \in G : gx = xg\}$$

(this last subgroup, called the *centralizer* of x in G, is denoted by $C_G(x)$).

The reader may check that the family of all orbits partitions X, for the relation $x \sim y$ on X, defined by $y = g \cdot x$ for some $g \in G$, is an equivalence relation on X whose equivalence classes are the orbits.

Theorem G.10. *If G acts on a set X and if $x \in X$, then*

$$|\mathcal{O}(x)| = [G : G_x] = |G|/|G_x|.$$

In particular, if G acts transitively on X, where $|X| = n$, then

$$|G| = n|G_x|.$$

Proof. Define $\varphi : \mathcal{O}(x) \rightarrow$ {the family of all cosets of G_x in G} by

$$\varphi(g \cdot x) = gG_x.$$

Now φ is well defined, for if $g \cdot x = h \cdot x$ (where $g, h \in G$), then $h^{-1}g \cdot x = x$, $h^{-1}g \in G_x$, and $gG_x = hG_x$. Reversing this argument shows that φ is an injection: if $\varphi(g \cdot x) = \varphi(h \cdot x)$, then $gG_x = hG_x$, $h^{-1}g \in G_x$, and $g \cdot x = h \cdot x$. Finally, φ is surjective, for a coset gG_x is $\varphi(g \cdot x)$. Hence, φ is a bijection.

If G acts transitively, then $\mathcal{O}(x) = X$ and $|\mathcal{O}(x)| = n = |X|$; hence $n = [G : G_x] = |G|/|G_x|$, and $|G| = n|G_x|$. $\bullet$

Corollary G.11. *If $x \in G$, then*

$$\text{the number of conjugates of } x = [G : C_G(x)].$$

Proof. This is the special case of G acting on itself by conjugation. $\bullet$

Lemma G.12. *If p is a prime not dividing m ($p \nmid m$) and if $k \geq 1$, then*

$$p \nmid \binom{p^k m}{p^k}.$$

Proof. Write the binomial coefficient as follows:

$$\binom{p^k m}{p^k} = \frac{p^k m(p^k m - 1) \cdots (p^k m - i) \cdots (p^k m - p^k + 1)}{p^k(p^k - 1) \cdots (p^k - i) \cdots (p^k - p^k + 1)}.$$

By Euclid's lemma, any factor p of the numerator (or of the denominator) arises from a factor of $p^k m - i$ (or of $p^k - i$). If $(m, p) = 1$ and $1 \leq i < p^k$, then $p^t \mid mp^k - i$ if and only if $p^t \mid i$. Therefore, the highest power of p dividing $p^k m - i$ is the same as the highest power of p dividing $p^k - i$ (because $p \nmid m$). Every factor of p upstairs is thus canceled by a factor of p downstairs, and hence the binomial coefficient has no factor p. $\bullet$

Theorem G.13 (Sylow). *If G is a group of order $p^k m$, where p is a prime not dividing m, then G contains a subgroup of order p^k.*

Proof. (Wielandt) If X is the family of all subsets of G of cardinality p^k, then Lemma G.12 shows that $p \nmid |X|$. Let G act on X by left translation: if $B \subset G$ and $|B| = p^k$, then

$$g \cdot B = \{gb : b \in B\}.$$

There is some orbit $\mathcal{O}(B)$ with $p \nmid |\mathcal{O}(B)|$ (otherwise p divides the cardinality of every orbit, hence p divides $|X|$). Choose such a subset $B \in X$. Now $|G|/|G_B| = [G : G_B] = |\mathcal{O}(B)|$ is prime to p; it follows that $|G_B| = p^k m' \geq p^k$ for some $m' \mid m$. On the other hand, if $b_0 \in B$ and $g \in G_B$, then $g b_0 \in g \cdot B = B$ (definition of stabilizer); moreover, if g and h are distinct elements of G_B, then $g b_0$ and $h b_0$ are distinct elements of B. Therefore $|G_B| \leq |B| = p^k$, and so G_B is a subgroup of order p^k.   •

Definition. If $|G| = p^k m$, where p is a prime not dividing m, then a subgroup of G of order p^k is called a *Sylow p-subgroup* of G.

One knows that any two Sylow p-subgroups of a group G are isomorphic (indeed, they are conjugate), and that there are exactly $1 + rp$ of them for some integer $r \geq 0$.

Corollary G.14 (Cauchy). *If p is a prime dividing $|G|$, then G contains an element of order p.*

Proof. Let H be a Sylow p-subgroup of G and choose $x \in H^{\#} = H - \{1\}$. By Lagrange's theorem, the order of x is p^t for some t. If $t = 1$, we are done; if $t > 1$, then it is easy to see that $x^{p^{t-1}}$ has order p.   •

Lemma G.15. *Every finite abelian group $G \neq \{1\}$ contains a subgroup of prime index.*

Proof. We first prove that if G has composite order rs, then G has a proper subgroup. Choose $x \in G$ with $x \neq 1$. If x has order $< rs$, then $\langle x \rangle$ is a proper subgroup; otherwise, x has order rs and $\langle x^r \rangle$ is a proper subgroup.

The proof of the lemma is by induction on the number k of (not necessarily distinct) prime factors of $|G|$. If $k = 1$, then G has prime order and $\{1\}$ has prime index. If $k > 1$, the first paragraph gives a proper subgroup H, necessarily normal (because G is abelian), and so the quotient group G/H is defined. By induction, G/H has a subgroup S^* of prime index, and the correspondence theorem gives a subgroup S of G of prime index.   •

Theorem G.16. *A group $G \neq \{1\}$ is solvable (it has a normal series with abelian factor groups) if and only if G has a normal series with factor groups of prime order.*

Proof. Sufficiency is obvious; we prove necessity by induction on $|G|$. Assume that

$$G = G_0 \supset G_1 \supset \cdots \supset G_n = \{1\}$$

is a normal series with G_i/G_{i+1} abelian for all i; we may further assume that $G \neq G_1$. By Lemma G.15, the abelian group G/G_1 has a (necessarily normal) subgroup S^* of prime index; the correspondence theorem gives an intermediate subgroup S ($G \supset S \supset G_1$) with S normal in G and with $[G : S] = [G/G_1 : S^*]$ prime. Now S is a solvable group (consider the normal series

$$S \supset G_1 \supset G_2 \supset \cdots \supset G_n = \{1\};$$

S/G_1 is abelian because it is a subgroup of the abelian group G/G_1), and induction provides a normal series of it with factor groups of prime order. •

Corollary G.17. *Every solvable group has a normal subgroup of prime index.*

Recall that the ***commutator*** of elements $x, y \in G$ is

$$[x, y] = xyx^{-1}y^{-1}.$$

The ***commutator subgroup*** G' of G is the subgroup generated by all the commutators (the product of two commutators may not be a commutator). Note that G' is a normal subgroup of G, for if $a \in G$, then

$$a[x, y]a^{-1} = [axa^{-1}, aya^{-1}];$$

moreover, G/G' is abelian.

Lemma G.18. *If H is a normal subgroup of G, then G/H is abelian if and only if $G' \subset H$.*

Proof. If G/H is abelian, then for all $x, y \in G$,

$$xyH = xHyH = yHxH = yxH,$$

and so $xyx^{-1}y^{-1} \in H$; it follows that $G' \subset H$ because every generator of G' lies in H. Conversely, if $G' \subset H$, then the third isomorphism theorem shows that G/H is a quotient group of the abelian group G/G', and hence it is abelian. •

Definition. The ***higher commutator subgroups*** are defined inductively:

$$G^{(0)} = G; \qquad G^{(i+1)} = G^{(i)\prime};$$

that is, $G^{(i+1)}$ is the commutator subgroup of $G^{(i)}$.

Lemma G.19. *A group G is solvable if and only if $G^{(n)} = \{1\}$ for some n.*

Proof. If G is solvable, then there is a normal series

$$G = G_0 \supset G_1 \supset \cdots \supset G_n = \{1\}$$

with each factor group G_i/G_{i+1} abelian. We prove, by induction on i, that $G_i \supset G^{(i)}$; this will give the result. If $i = 0$, then $G_i = G_0 = G$. Assume, by induction, that $G_i \supset G^{(i)}$; then $G_i' \supset G^{(i)'} = G^{(i+1)}$. But G_i/G_{i+1} abelian implies $G_{i+1} \supset G_i'$, by Lemma G.18, and so $G_{i+1} \supset G^{(i+1)}$.

Conversely, if $G^{(n)} = \{1\}$ (of course, $G^{(1)} = G'$), then

$$G = G^{(0)} \supset G^{(1)} \supset G^{(2)} \supset \cdots \supset G^{(n)} = \{1\}$$

is a normal series with abelian factor groups; hence G is solvable.   •

Theorem G.20. *If G is a solvable group, then every subgroup and every quotient group of G is also solvable.*

Proof. If H is a subgroup of G, then it is easy to prove by induction that $H^{(i)} \subset G^{(i)}$ for all i. Hence, $G^{(n)} = \{1\}$ implies $H^{(n)} = \{1\}$ and H is solvable.

If $\varphi : G \to K$ is a surjective homomorphism, then $\varphi(G') = K'$: if $uvu^{-1}v^{-1}$ is a commutator in K, choose $x, y \in G$ with $\varphi(x) = u$ and $\varphi(y) = v$; then $\varphi(xyx^{-1}y^{-1}) = uvu^{-1}v^{-1}$. One proves easily, by induction, that $\varphi(G^{(i)}) = K^{(i)}$ for all i. Hence, if G is solvable, then $G^{(n)} = \{1\}$ for some n and $K^{(n)} = \{1\}$; therefore K is solvable. Now take $K = G/N$, where N is any normal subgroup of G, and take φ to be the natural map $G \to G/N$.   •

Theorem G.21. *Let G be a group with normal subgroup H. If H and G/H are solvable groups, then G is solvable.*

Proof. Let

$$G/H = G^* = G_0^* \supset G_1^* \supset \cdots \supset G_m^* = \{1\}$$

be a normal series with abelian factor groups. By the correspondence theorem, there is a series

$$G = G_0 \supset G_1 \supset \cdots \supset G_m = H$$

with each G_i normal in G_{i-1} and with abelian factor groups. Since H is solvable, there is a normal series

$$H = H_0 \supset H_1 \supset \cdots \supset H_n = \{1\}$$

with abelian factor groups. Splicing these two series together gives a normal series for G with abelian factor groups.   •

One can also prove this result using the criterion in Lemma G.19.

Definition. The *center* of a group G is

$$Z(G) = \{g \in G : gx = xg \quad \text{for all } x \in G\}.$$

It is easy to see that $Z(G)$ is an abelian normal subgroup of G.

It is also easy to prove that $g \in Z(G)$ if and only if the conjugacy class of g is $\{g\}$, so that $|Z(G)|$ is the number of conjugacy classes of cardinality 1.

There are groups G with $Z(G) = \{1\}$; for example, $Z(S_3) = \{1\}$.

Lemma G.22. *If p is a prime and $G \neq \{1\}$ is a p-group, then $Z(G) \neq \{1\}$.*

Proof. Partition G into its conjugacy classes: using our remark above about conjugacy classes of cardinality 1, there is a disjoint union

$$G = Z(G) \cup C_1 \cup \cdots \cup C_t,$$

where the C_i are the conjugacy classes of cardinality larger than 1. If we choose $x_i \in C_i$, then Corollary G.11 gives

$$|G| = |Z(G)| + \sum [G : C_G(x_i)].$$

By Lagrange's theorem, p divides $[G : C_G(x_i)]$ for all i (if $x_i \notin Z(G)$, then $C_G(x_i) \neq G$ and $[G : C_G(x_i)] \neq 1$), and so p divides $|Z(G)|$.   •

Theorem G.23. *Every p-group G is solvable, and hence it has a normal subgroup of index p if $G \neq \{1\}$.*

Proof. We prove that G is solvable by induction on $|G|$. If $|G| \neq 1$, then $Z(G) \neq \{1\}$, by Lemma G.22. If $Z(G) = G$, then G is abelian, hence solvable. If $Z(G) \neq G$, then $G/Z(G)$ is a p-group of order $< |G|$, hence it is solvable, by induction. Since $Z(G)$ is solvable, being abelian, Theorem G.21 shows that G is solvable.

As G is solvable, the second statement follows from Corollary G.17.   •

Let us pass from abstract groups to permutation groups; Cayley's theorem shows that this is no loss in generality.

Recall that S_X, the symmetric group on a set X, is the set of all permutations (bijections) of X under composition. If $X = \{x_1, \ldots, x_n\}$, then there is an isomorphism $S_X \to S_n$ (namely, $\alpha \mapsto \theta\alpha\theta^{-1}$, where $\theta(x_i) = i$) and one usually identifies these two groups.

Theorem G.24 (Cayley). *Every group G of order n is (isomorphic to) a subgroup of S_n.*

Proof. If $a \in G$, then the function $\lambda_a : G \to G$, defined by $x \mapsto ax$, is a bijection, for its inverse is $\lambda_{a^{-1}} : x \mapsto a^{-1}x$; hence $\lambda_a \in S_G$ (of course, $S_G \cong S_n$). Define $\lambda : G \to S_G$ by $a \mapsto \lambda_a$. It remains to prove that λ is an injective homomorphism.

If $a, b \in G$ are distinct, then $\lambda_a \neq \lambda_b$ (because these two functions have different values on $1 \in G$). Finally, λ is a homomorphism:

$$\lambda_a \lambda_b : x \mapsto bx \mapsto a(bx)$$

and

$$\lambda_{ab} : x \mapsto (ab)x,$$

so the associative law implies $\lambda_{ab} = \lambda_a \lambda_b$, as desired.   •

Lemma G.25. *The alternating group A_n is generated by the 3-cycles.*

Proof. If $\alpha \in A_n$, then $\alpha = \tau_1 \cdots \tau_m$, where each τ_i is a transposition and m is even; hence

$$\alpha = (\tau_1 \tau_2)(\tau_3 \tau_4) \cdots (\tau_{m-1} \tau_m).$$

If τ_{2k-1} and τ_{2k} are not disjoint, then their product is a 3-cycle: $\tau_{2k-1} \tau_{2k} = (ab)(ac) = (acb)$;[18] if they are disjoint, then

$$\tau_{2k-1} \tau_{2k} = (ab)(cd) = (ab)(bc)(bc)(cd) = (bca)(cdb).$$

Therefore α is a product of 3-cycles.   •

Lemma G.26. *The commutator subgroup of S_n is A_n.*

Proof. Since S_n/A_n is abelian (it has order 2), Lemma G.18 gives $S_n' \subset A_n$. Since A_n is generated by the 3-cycles, it suffices to prove every $\sigma = (ijk)$ is a commutator. Since σ has order 3, $\sigma = \sigma^4 = (\sigma^2)^2$. But

$$\sigma^2 = (ikj) = (ij)(ik),$$

---

[18]We multiply permutations from right to left:

$$(\sigma \tau)a = \sigma(\tau(a))$$

because we are composing functions: that is, $\sigma \tau : a \mapsto \tau a \mapsto \sigma(\tau a)$. In particular, $(ab)(ac) = (acb)$ because

$$(ab)(ac) : a \mapsto c \mapsto c; \quad b \mapsto b \mapsto a; \quad c \mapsto a \mapsto b.$$

so that
$$\sigma = \sigma^4 = (ij)(ik)(ij)(ik);$$
this is a commutator because $(ij) = (ij)^{-1}$ and $(ik) = (ik)^{-1}$. •

Lemma G.27. *If* $\gamma = (i_0, i_1, \dots, i_{k-1})$ *is a k-cycle in* S_n *and* $\alpha \in S_n$, *then* $\alpha\gamma\alpha^{-1}$ *is also a k-cycle; indeed,*
$$\alpha\gamma\alpha^{-1} = (\alpha i_0, \alpha i_1, \dots, \alpha i_{k-1}).$$

Conversely, if $\gamma' = (i'_0, i'_1, \dots, i'_{k-1})$ *is another k-cycle, then there exists* $\alpha \in S_n$ *with* $\gamma' = \alpha\gamma\alpha^{-1}$.

Proof. If $\ell \neq \alpha i_j, 0 \leq j \leq k-1$, then $\alpha^{-1}\ell \neq i_j$ and so $\gamma(\alpha^{-1}\ell) = \alpha^{-1}\ell$; therefore $\alpha\gamma\alpha^{-1}: \ell \mapsto \alpha^{-1}\ell \mapsto \alpha^{-1}\ell \mapsto \ell$; that is, $\alpha\gamma\alpha^{-1}$ fixes ℓ. If $\ell = \alpha i_j$, then $\alpha\gamma\alpha^{-1} : \ell = \alpha i_j \mapsto i_j \mapsto i_{j+1} \mapsto \alpha i_{j+1}$ (read subscripts mod k). Hence $\alpha\gamma\alpha^{-1}$ and $(\alpha i_0, \alpha i_1, \dots, \alpha i_{k-1})$ are equal.

Conversely, given γ and γ', choose a permutation α with $\alpha i_j = i'_j$ for all j. Then the first part of the proof shows that $\gamma' = \alpha\gamma\alpha^{-1}$. •

Remark. The same technique proves the lemma with γ a cycle replaced by γ a product of disjoint cycles.

Lemma G.28. *If H is a subgroup of a group G of index 2, then H is a normal subgroup of G.*

Proof. If $a \in G$ and $a \notin H$, then $aH \cap H = \emptyset$ and, by hypothesis, $aH \cup H = G$; hence aH is the complement of H. Since $Ha \cap H = \emptyset$, it follows that $Ha \subset aH$; that is, after multiplying on the right by a^{-1},
$$H \subset aHa^{-1}.$$
This inclusion holds for every $a \in G$, so we may replace a by a^{-1} to obtain $H \subset a^{-1}Ha$; that is, $aHa^{-1} \subset H$. Therefore, H is a normal subgroup of G. •

Theorem G.29. *The alternating group* A_n *is the only subgroup of* S_n *having index 2.*

Proof. If $[S_n : H] = 2$, then H is normal in S_n, by Lemma G.28, and Lemma G.18 gives $A_n = S'_n \subset H$ (for S_n/H has order 2, hence is abelian). But $|A_n| = n!/2 = |H|$, and so $H = A_n$. •

We are going to prove that A_5 is a simple group.

Lemma G.30. (i) *There are* 20 3-*cycles in* S_5, *and they are all conjugate in* S_5.

(ii) *All* 3-*cycles are conjugate in* A_5.

Proof. (i) The number of 3-cycles (abc) is $5 \times 4 \times 3/3 = 20$ (one divides by 3 because $(abc) = (bca) = (cab)$). The conjugacy of any two 3-cycles follows at once from Lemma G.27.

(ii) Given 3-cycles γ, γ', one must find an *even* permutation α with $\gamma' = \alpha\gamma\alpha^{-1}$. This can be done directly, but it involves consideration of various cases; here is another proof.

If $\alpha = (123)$ and $C_S(\alpha)$ is the centralizer of α in S_5, then Corollary G.11 gives $20 = [S_5 : C_S(\alpha)]$; hence $|C_S(\alpha)| = 6$. But we can exhibit the six elements that commute with α:

$$1, \cdot \; \alpha, \quad \alpha^2, \quad (45), \quad (45)\alpha, \quad (45)\alpha^2.$$

Only the first three of these are even permutations, and so $|C_A(\alpha)| = 3$, where $C_A(\alpha)$ is the centralizer of α in A_5. By Corollary G.11, the number of conjugates of α in A_5 is $[A_5 : C_A(\alpha)] = |A_5|/|C_A(\alpha)| = 60/3 = 20$. Therefore, all 3-cycles are conjugate to $\alpha = (123)$ in A_5. $\bullet$

Theorem G.31. A_5 *is a simple group*.

Proof. If $H \neq \{1\}$ is a normal subgroup of A_5 and if $\sigma \in H$, then every conjugate of σ in A_5 also lies in H. In particular, if H contains a 3-cycle, then it contains all 3-cycles, by Lemma G.30(ii); but then $H = A_5$, by Lemma G.25.

Let $\sigma \in H, \sigma \neq 1$. After a harmless relabeling, we may assume either $\sigma = (123)$, $\sigma = (12)(34)$, or $\sigma = (12345)$ (these are the only possible cycle structures of (even) permutations in A_5). If $\sigma = (123)$, then $H = A_5$, as we have noted above. If $\sigma = (12)(34)$, define $\tau = (12)(35)$; then

$$\tau\sigma\tau^{-1} = (\tau 1 \, \tau 2)(\tau 3 \, \tau 4) = (12)(45)$$

and

$$\tau\sigma\tau^{-1}\sigma^{-1} = (354) \in H.$$

Finally, if $\sigma = (12345)$, define $\tau = (132)$; then

$$\sigma\tau^{-1}\sigma^{-1} = (\sigma 1 \, \sigma 2 \, \sigma 3) = (234)$$

and

$$\tau\sigma\tau^{-1}\sigma^{-1} = (134).$$

In each case, H must contain a 3-cycle. Therefore, A_5 contains no proper normal subgroups $\neq \{1\}$ and hence it is simple. $\bullet$

One can prove, by induction, that A_n is simple for all $n \geq 5$.
The next counting lemma is useful.

Lemma G.32. *If A and B are subgroups of a finite group G, then*

$$|A \cap B||AB| = |A||B|,$$

where AB is the subset $\{ab : a \in A \text{ and } b \in B\}$.

Proof. We are going to use the following fact. If X and Y are finite sets and $\varphi : X \to Y$ is a surjection for which $|\varphi^{-1}(y)| = |\varphi^{-1}(y')|$ for all $y, y' \in Y$, then $|Y| = |X|/|\varphi^{-1}(y)|$.

Define $\varphi : A \times B \to AB$ by $(a, b) \mapsto ab$; of course, φ is a surjection. We claim that

$$\varphi^{-1}(ab) = \{(ac, c^{-1}b) : c \in A \cap B\}.$$

It is clear that $(ac, c^{-1}b) \in \varphi^{-1}(ab)$. Conversely, if $(\alpha, \beta) \in \varphi^{-1}(ab)$, then $ab = \alpha\beta$, where $\alpha \in A$ and $\beta \in B$. Hence, $\alpha^{-1}a = \beta b^{-1} \in A \cap B$; if c is their common value, then

$$(\alpha, \beta) = (ac, c^{-1}b).$$

Therefore, $|\varphi^{-1}(ab)| = |A \cap B|$ and $|AB| = |A \times B|/|A \cap B|$. •

Corollary G.33. *The only normal subgroups of S_5 are $\{1\}$, A_5, and S_5.*

Proof. Let $H \neq \{1\}$ be a normal subgroup of S_5. The second isomorphism theorem gives $H \cap A_5$ a normal subgroup of A_5; as A_5 is a simple group, either $H \cap A_5 = A_5$ or $H \cap A_5 = \{1\}$. In the first case, $A_5 \subset H$ and $H = A_5$ or $H = S_5$. In the second case, there is $h \in H$ with $h \notin A_5$, so that $HA_5 = S_5$. Since $H \cap A_5 = \{1\}$, Lemma G.32 gives $|H| = |S_5|/|A_5| = 2$. If $h \in H$, $h \neq 1$, then $h = (ab)$ (the only other elements of order 2 have the form $(ab)(cd)$, and they are even permutations). It is easy to find a conjugate distinct from h, and this contradicts the normality of H. •

Theorem G.34. *S_n is solvable for $n \leq 4$, but it is not solvable for $n \geq 5$.*

Proof. If $m < n$, then S_m is (isomorphic to) a subgroup of S_n. Since every subgroup of a solvable group is itself solvable (Theorem G.20), it suffices to show that S_4 is solvable and S_5 is not solvable.

Here is a normal series of S_4 that has abelian factor groups:

$$S_4 \supset A_4 \supset V \supset \{1\},$$

where V is the four group (the factor groups have orders 2, 3, 4, respectively, hence are abelian).

Were S_5 solvable, then its subgroup A_5 would also be solvable. Since A_5 is simple, its only normal series is $A_5 \supset \{1\}$, and the (only) factor group is the nonabelian group $A_5/\{1\} \cong A_5$. •

We now discuss Exercise 106, the group theoretic basis of the computation of the Galois groups of irreducible quartic polynomials over $\mathbb{Q}$.

First of all, we list the subgroups G of S_4 whose order is a multiple of 4. If $|G| = 4$, then the only abstract groups G are $\mathbb{Z}_4$ and $\mathbb{Z}_2 \times \mathbb{Z}_2$, and both occur as subgroups of S_4 (in particular, $V \cong \mathbb{Z}_2 \times \mathbb{Z}_2$). There is a subgroup of order 8 isomorphic to the dihedral group D_8, namely, the symmetries of a square regarded as permutations of the 4 corners; since a subgroup of order 8 is a Sylow 2-subgroup of S_4, all subgroups of order 8 are isomorphic to D_8. Theorem G.29 shows that A_4 is the only subgroup of order 12 and, of course, S_4 itself is the only subgroup of order 24.

If $G \subset S_4$ and V is the four group (which is a normal subgroup of S_4), then the second isomorphism theorem gives $G \cap V \lhd G$ and

$$G/G \cap V \cong GV/V \subset S_4/V.$$

Define

$$m = |G/G \cap V|;$$

it follows that m is a divisor of $[S_4 : V] = 24/4 = 6$ ($S_4/V \cong S_3$, but we do not need this fact.)

Theorem G.35 (Exercise 106). *Let* $G \subset S_4$ *have order a multiple of* 4 *and let* $m = |G/G \cap V|$.
   (i) *If* $m = 6$, *then* $G = S_4$;
  (ii) *if* $m = 3$, *then* $G = A_4$;
 (iii) *if* $m = 1$, *then* $G = V$;
 (iv) *if* $m = 2$, *then* $G \cong D_8$ *or* $\mathbb{Z}_4$ *or* V.

Proof. If $m = 6$ or 3, then $|G| \geq 12$ ($|G|$ is divisible by 3 and, by hypothesis, 4). By Theorem G.29, A_4 is the only subgroup of S_4 of order 12, and so $A_4 \subset G$ in either case. But $V \subset A_4$. It follows easily that $m = 6$ forces $G = S_4$ and $m = 3$ forces $G = A_4$.

If $m = 1$, then $G = G \cap V$ and $G \subset V$; since $|G|$ is a multiple of 4, it follows that $G = V$.

If $m = 2$, then $|G| = 2|G \cap V|$; since $|V| = 4$, we have $|G \cap V| = 1$, 2, or 4. We cannot have $|G \cap V| = 1$ lest $|G| = 2$, which is not a multiple of 4. If $|G \cap V| = 4$, then $|G| = 8$ and $G \cong D_8$ (as we remarked above, D_8 is a Sylow 2-subgroup). If $|G \cap V| = 2$, then $|G| = 4$ and $G \cong \mathbb{Z}_4$ or V (these are the only abstract groups of order 4). •

The possibility $m = 2$ and $G \cong V$ can occur. Let G be the following isomorphic copy of V in S_4:

$$G = \{1, (12)(34), (12), (34)\}.$$

Note that $G \cap V = \{1, (12)(34)\}$ and $m = |G/G \cap V| = 4/2 = 2$. This group G does not act transitively on $\{1, 2, 3, 4\}$ because, for example, there is no $g \in G$ with $g(1) = 3$. Exercise 107 invokes the extra hypothesis of G acting transitively to eliminate the case $G \cong V$ from the list of candidates for G when $m = 2$.

Lemma G.36. *If G is a group and H is a subgroup of index n, then there is a homomorphism $\varphi : G \to S_n$ with $\ker \varphi \subset H$.*

Proof. Let X be the family of all cosets of H in G; since $|X| = n$, it is easy to see that $S_X \cong S_n$ (where S_X is the group of all permutations of X). For $g \in G$, define $\varphi(g) : X \to X$ by $\varphi(g) : aH \mapsto gaH$ (where $a \in G$); note that $\varphi(g)$ is a bijection, for its inverse is $\varphi(g^{-1})$. To see that φ is a homomorphism, compute:

$$\varphi(gg') : aH \mapsto (gg')aH;$$
$$\varphi(g)\varphi(g') : aH \mapsto g'aH \mapsto g(g'aH).$$

If $\varphi(g)$ is the identity on X, then $\varphi(g) : aH \mapsto aH$ for all $a \in G$; in particular, $\varphi(g) : H \mapsto H$, so that $gH = H$ and $g \in H$. •

Theorem G.37. *A_6 has no subgroups of prime index.*

Proof. Now A_6 is a simple group of order $360 = 2^3 \cdot 3^2 \cdot 5$ (in fact, A_n is a simple group of order $\frac{1}{2}n!$ for all $n \geq 5$). If H is a subgroup of prime index, then $[A_6 : H] = 2, 3,$ or 5. By Lemma G.36, there is a homomorphism $\varphi : A_6 \to S_n$, where $n = 2, 3,$ or 5, with $\ker \varphi \subset H$; in particular, $\ker \varphi$ is a normal subgroup of A_6 with $\ker \varphi \neq A_6$. Since A_6 is simple, $\ker \varphi = \{1\}$ and φ is an injection. But this is impossible because $|S_5| = 120 < 360$. •

Lemma G.38. *S_5 has no subgroups of order 30 or of order 40.*

Proof. If H is a subgroup of order 30, then H has index $[S_5 : H] = 120/30 = 4$. Lemma G.36 gives a homomorphism $\varphi : S_5 \to S_4$ with

$\ker \varphi \subset H$. But $\ker \varphi$ is a normal subgroup of S_5, and so its order must be 1, 60, or 120 (Corollary G.33). Since $|H| = 30$, it follows that $\ker \varphi = \{1\}$, and S_5 is isomorphic to a subgroup of S_4, a contradiction. A similar argument shows that S_5 has no subgroup of index 3.   •

Theorem G.39. *If α is a 5-cycle in S_5 and τ is a transposition in S_5, then* $\langle \alpha, \tau \rangle = S_5$.

Proof. Let $H = \langle \alpha, \tau \rangle$ be the subgroup generated by α and τ. We may assume that $\alpha = (1\,2\,3\,4\,5)$ and $\tau = (1\,i)$. Now some power of α, say, α^k carries i into 1, so that Lemma G.27 gives $\alpha^k (1\,i) \alpha^{-k} = (j\,1)$ for some j (actually, $j = \alpha^k 1$). Note that $i \neq j$ because α^k does not commute with $(1i)$. But $(1i)(1j) = (1\,j\,i)$, an element of order 3. The order of H is thus divisible by 2, 3, and 5, hence $|H| \geq 30$. By Lemma G.38, $|H| = 60$ or 120. If $|H| = 60$, then $H = A_5$, by Theorem G.29; but $H \neq A_5$ because $\tau \in H$ is an odd permutation. Therefore $H = S_5$.   •

A more computational proof shows first that every transposition can be obtained from α and τ, and then that S_5 is generated by the transpositions.

Theorem G.40. *A subgroup H of S_5 is solvable if and only if $|H| \leq 24$.*

Proof. We leave to the reader the fact that every group of order ≤ 24 is solvable (whether or not it is a subgroup of S_5; indeed, every group of order < 60 is solvable).

Since $|S_5| = 120$, the only divisors of $|S_5|$ larger than 24 are 30, 40, 60, and 120. Now S_5 itself is not solvable, by Theorem G.34; also, A_5 is the only subgroup of order 60 (Theorem G.29), and it is not solvable because it is simple and not abelian (Theorem G.31). Lemma G.38 completes the proof.   •

Theorem G.40 is used in Exercise 111. It is implicit in the second part of this exercise that S_5 does have a subgroup of order 20; the **normalizer** of a Sylow 5-subgroup is such a subgroup, where the normalizer $N_G(P)$ of a subgroup P of G is defined as:

$$N_G(P) = \{g \in G : g P g^{-1} = P\}.$$

Of course, S_5 does have a solvable subgroup of order 24, namely, S_4.

Appendix C
Ruler-Compass Constructions

We are going to show that the classical Greek problems: squaring the circle, duplicating the cube, and trisecting an angle, are impossible to solve. As we shall see, the discussion uses only elementary field theory; no Galois theory is required.

It is clear one that can trisect a 60° angle with a protractor (or any other device than can measure an angle); after all, one can divide any number by 3. Therefore, it is essential to state the problems carefully and to agree on certain ground rules. The Greek problems specify that only two tools are allowed, and each must be used in only one way. Let P and Q be points in the plane; we denote the line segment with endpoints P and Q by PQ, and we denote the length of this segment by $|PQ|$. A *ruler* (or *straight-edge*) is a tool that can draw the line $L(P, Q)$ determined by P and Q; a *compass* is a tool that draws the circle with radius $|PQ|$ and center either P or Q; denote these circles by $C(P; Q)$ or $C(Q; P)$, respectively. Since every construction has only a finite number of steps, we shall be able to define "constructible" points inductively.

Given the plane, we establish a coordinate system by first choosing two distinct points, A and $\overline{A}$; call the line they determine the *x-axis*. Use a compass to draw the two circles $C(A; \overline{A})$ and $C(\overline{A}; A)$ of radius $|A\overline{A}|$ with centers A and $\overline{A}$, respectively. These two circles intersect in two points; the line they determine is called the *y- axis*; it is the perpendicular bisector of $A\overline{A}$, and it intersects the *x*-axis in a point O, called the *origin*. We define the distance $|OA|$ to be 1. We have introduced coordinates in the plane; in particular, $A = (1, 0)$ and $\overline{A} = (-1, 0)$.

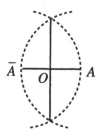

Figure 5

Informally, one constructs a new point T from (not necessarily distinct) old points P, Q, R, and S by using the first pair P, Q to draw a line or circle, the second pair R, S to draw a line or circle, and then obtaining T as one of the points of intersection of the two drawn lines, the drawn line and the drawn circle, or the two drawn circles. More generally, a point is called constructible if it is obtained from A and $\overline{A}$ by a finite number of such steps. Given a pair of constructible points, we do *not* assert that every point on the drawn line or the drawn circles they determine is constructible.

Here is the formal discussion.

Definition. Let E, F, G, and H be (not necessarily distinct) points in the plane. A point Z is *constructible from* E, F, G, and H if either

(i) $Z \in L(E, F) \cap L(G, H)$, where $L(E, F) \neq L(G, H)$;

(ii) $Z \in L(E, F) \cap C(G; H)$;

(iii) $Z \in C(E; F) \cap C(G; H)$, where $C(E; F) \neq C(G; H)$.

A point Z is *constructible* if $Z = A$ or $Z = \overline{A}$ or if there are points $P_1, \ldots, P_n$ with $Z = P_n$ so that, for all $j \geq 1$, the point P_{j+1} is constructible from points in $\{A, \overline{A}, P_1, \ldots, P_j\}$.

Example 38. Let us show that $Z = (0, 1)$ is constructible. We have seen above that the origin $P_1 = O$ is constructible. The points $P_2 = (0, \sqrt{3})$ and $P_3 = (0, -\sqrt{3})$ are constructible, for both lie in $C(A; \overline{A}) \cap C(\overline{A}; A)$, and so the y-axis $L(P_2, P_3)$ can be drawn. Finally,

$$Z = (0, 1) \in L(P_2, P_3) \cap C(O; A).$$

In our discussion, we shall freely use any standard result of euclidean geometry. For example, every angle can be bisected with ruler and compass; i.e., if $(\cos \theta, \sin \theta)$ is constructible, then so is $(\cos \theta/2, \sin \theta/2)$.

Definition. A complex number $z = x + iy$ is *constructible* if the point (x, y) is a constructible point.

Example 38 shows that the numbers $1, -1, 0, i\sqrt{3}, -i\sqrt{3}$, and i are constructible numbers.

Lemma R.1. *A complex number $z = x + iy$ is constructible if and only if its real part x and its imaginary part y are constructible.*

Proof. If z is constructible, then a standard euclidean construction draws the vertical line L through (x, y) which is parallel to the y-axis. It follows that x is constructible, for the point $(x, 0)$ is constructible, being the intersection of L and the x-axis. Similarly, the point $(0, y)$ is the intersection of the y-axis and a line through (x, y) which is parallel to the x-axis. It follows that $P = (y, 0)$ is constructible, for it is an intersection point of the x-axis and $C(O; P)$. Hence, y is a constructible number.

Conversely, assume that x and y are constructible numbers; that is, $Q = (x, 0)$ and $P = (y, 0)$ are constructible points. The point $(0, y)$ is constructible, being the intersection of the y-axis and $C(O; P)$. One can draw the vertical line through $(x, 0)$ as well as the horizontal line through $(0, y)$, and (x, y) is the intersection of these lines. Therefore, (x, y) is a constructible point, and so $z = x + iy$ is a constructible number. ●

Definition. We denote by K the subset of $\mathbb{C}$ consisting of all the ***constructible numbers***.

Lemma R.2.
  (i) *If $K \cap \mathbb{R}$ is a subfield of $\mathbb{R}$, then K is a subfield of $\mathbb{C}$.*
  (ii) *If $K \cap \mathbb{R}$ is a subfield of $\mathbb{R}$ and if $\sqrt{a} \in K$ whenever $a \in K \cap \mathbb{R}$ is positive, then K is closed under square roots.*

Proof. (i) If $z = a + ib$ and $w = c + id$ are constructible, then $a, b, c, d \in K \cap \mathbb{R}$, by Lemma R.1. Hence, $a + c, b + d \in K \cap \mathbb{R}$, because $K \cap \mathbb{R}$ is a subfield, and so $(a + c) + i(b + d) \in K$, by Lemma R.1. Similarly, $zw = (ac - bd) + i(ad + bc) \in K$. If $z \neq 0$, then $z^{-1} = (a/z\bar{z}) - i(b/z\bar{z})$. Now $a, b \in K \cap \mathbb{R}$, by Lemma R.1, so that $z\bar{z} = a^2 + b^2 \in K \cap \mathbb{R}$, because $K \cap \mathbb{R}$ is a subfield of $\mathbb{C}$. Therefore, $z^{-1} \in K$.

(ii) If $z = a + ib \in K$, then $a, b \in K \cap \mathbb{R}$, by Lemma R.1, and so $r^2 = a^2 + b^2 \in K \cap \mathbb{R}$, as in part (i). Since r^2 is non-negative, the hypothesis gives $r \in K \cap \mathbb{R}$ and $\sqrt{r} \in K \cap \mathbb{R}$. Now $z = re^{i\theta}$, so that $e^{i\theta} = r^{-1}z \in K$, because K is a subfield of $\mathbb{C}$. That every angle can be bisected gives $e^{i\theta/2} \in K$, and so $\sqrt{z} = \sqrt{r}e^{i\theta/2} \in K$, as desired. ●

Theorem R.3. *The set of all constructible numbers K is a subfield of $\mathbb{C}$ that is closed under square roots and complex conjugation.*

Proof. For the first two statements, it suffices to prove that the properties of $K \cap \mathbb{R}$ in Lemma R.2 do hold. Let a and b be constructible reals.

(i) $-a$ *is constructible.*

If $P = (a, 0)$ is a constructible point, then $(-a, 0)$ is the other intersection of the x-axis and $C(O; P)$.

(ii) $a + b$ *is constructible.*

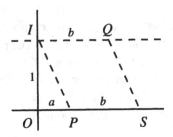

Figure 6

Let $I = (0, 1)$, $P = (a, 0)$ and $Q = (b, 1)$. Now Q is constructible: it is the intersection of the horizontal line through I and the vertical line through $(b, 0)$ [the latter point is constructible, by hypothesis]. The line through Q parallel to IP intersects the x-axis in $S = (a + b, 0)$, as desired. Although Figure 6 is drawn with a, b positive, it is clear that this construction works for any choice of signs of a, b.

(iii) ab *is constructible.*

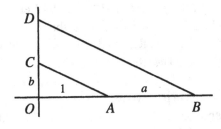

Figure 7

By (i), we may assume that both a and b are positive. In Figure 7, $A = (1, 0)$, $B = (1 + a, 0)$, and $C = (0, b)$. Define D to be the intersection of the y-axis and the line through B parallel to AC. Since the triangles OAC and OBD are similar,

$$|OB|/|OA| = |OD|/|OC|;$$

hence $(a+1)/1 = (b+|CD|)/b$, and $|CD| = ab$. Therefore, $b + ab$ is constructible. Since $-b$ is constructible, by (i), we have $ab = (b+ab) - b$ constructible, by (ii).

(iv) *If $a \neq 0$, then a^{-1} is constructible.*

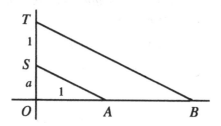

Figure 8

Let $A = (1, 0)$, $S = (0, a)$, and $T = (0, 1+a)$. Define B as the intersection of the x-axis and the line through T parallel to AS; thus, $B = (1 + u, 0)$ for some u. Similarity of the triangles OSA and OTB gives

$$|OT|/|OS| = |OB|/|OA|.$$

Hence, $(1 + a)/a = (1 + u)/1$, and so $u = a^{-1}$. Therefore, $1 + a^{-1}$ is constructible, and so $(1 + a^{-1}) - 1 = a^{-1}$ is constructible.

(v) *If $a \geq 0$, then $\sqrt{a}$ is constructible.*

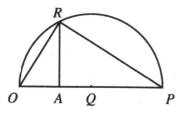

Figure 9

Let $A = (1, 0)$ and $P = (1 + a, 0)$; construct Q, the midpoint of OP. Define R as the intersection of the circle $C(Q; O)$ with the vertical line through A. The (right) triangles AOR and ARP are similar, so that

$$|OA|/|AR| = |AR|/|AP|,$$

and so $|AR| = \sqrt{a}$.

(vi) *If $z = a + ib \in K$, then $\bar{z} = a - ib$ is constructible.*

By Lemma R.2, K is a subfield of $\mathbb{C}$. Now $a, b \in K$, by Lemma R.1, and $i \in K$, by Example 38. Therefore, $-bi \in K$, and so $a - ib \in K$.   •

Corollary R.4. *If a, b, c are constructible, then the roots of the quadratic $ax^2 + bx + c$ are also constructible.*

Proof. This follows from the theorem and the quadratic formula.   •

We now consider subfields of $\mathbb{C}$ to enable us to prove an inductive step in the upcoming theorem.

Lemma R.5. *Let F be a subfield of $\mathbb{C}$ that contains i and that is closed under complex conjugation. Let $z = a + ib$, $w = c + id \in F$, and let $P = (a, b)$ and $Q = (c, d)$.*

(i) *If $a + ib \in F$, then $a \in F$ and $b \in F$.*

(ii) *If the equation of $L(P, Q)$ is $y = mx + q$, where $m, q \in \mathbb{R}$, then $m, q \in F$.*

(iii) *If the equation of $C(P; Q)$ is $(x - a)^2 + (y - b)^2 = r^2$, where $a, b, r \in \mathbb{R}$, then $r^2 \in F$.*

Proof. (i) If $z = a + ib \in F$, then $a = \frac{1}{2}(z + \bar{z}) \in F$ and $ib = \frac{1}{2}(z - \bar{z}) \in F$; since we are assuming $i \in F$, we have $b \in F$.

(ii) If $L(P, Q)$ is not vertical, its equation is $y - b = m(x - a)$. Now $m = (d - b)/(c - a) \in F$, since $a, b, c, d \in F$, and so $q = -ma + b \in F$.

(iii) The circle $C(P; Q)$ has equation $(x - a)^2 + (y - b)^2 = r^2$, and $r^2 = (c - a)^2 + (d - b)^2 \in F$.   •

Lemma R.6. *Let F be a subfield of $\mathbb{C}$ that contains i and that is closed under complex conjugation. Let P, Q, R, S be points whose coordinates lie in F, and let $\alpha = u + iv \in \mathbb{C}$. If either*

$$\alpha \; \in \; L(P, Q) \cap L(R, S), \; \text{where } L(P, Q) \neq L(R, S),$$

$$\alpha \; \in \; L(P, Q) \cap C(R; S),$$

or

$$\alpha \; \in \; C(P; Q) \cap C(R, S), \; \text{where } C(P; Q) \neq C(R; S),$$

then $[F(\alpha) : F] \leq 2$.

Proof. If $L(P, Q)$ is not vertical, then Lemma R.5(ii) says that $L(P, Q)$ has equation $y = mx + b$, where $m, b \in F$. If $L(P, Q)$ is vertical, then its equation is $x = b$ because $P = (a, b) \in L(P, Q)$, and so $b \in F$, by

Lemma R.5(i). Similarly, $L(R, S)$ has equation $y = nx + c$ or $x = c$, where $m, b, n, c \in F$. Since these lines are not parallel, one can solve the pair of linear equations for (u, v), the coordinates of $\alpha \in L(P, Q) \cap L(R, S)$, and they also lie in F. In this case, therefore, $[F(\alpha) : F] = 1$.

Let $L(P, Q)$ have equation $y = mx + q$ or $x = q$, and let $C(R; S)$ have equation $(x - c)^2 + (y - d)^2 = r^2$; by Lemma R.5, we have $m, q, r^2 \in F$. Since $\alpha = u + iv \in L(P, Q) \cap C(R; S)$,

$$\begin{aligned} r^2 &= (u - c)^2 + (v - d)^2 \\ &= (u - c)^2 + (mu + q - d)^2, \end{aligned}$$

so that u is a root of a quadratic polynomial with coefficients in $F \cap \mathbb{R}$. Hence, $[F(u) : F] \leq 2$. Since $v = mu + q$, we have $v \in F(u)$, and, since $i \in F$, we have $\alpha \in F(u)$. Therefore, $\alpha = u + iv \in F(u)$, and so $[F(\alpha) : F] \leq 2$.

Let $C(P; Q)$ have equation $(x - a)^2 + (y - b)^2 = r^2$, and let $C(R; S)$ have equation $(x - c)^2 + (y - d)^2 = s^2$. By Lemma R.5, we have $r^2, s^2 \in F \cap \mathbb{R}$. Since $\alpha \in C(P; Q) \cap C(R; S)$, there are equations

$$(u - a)^2 + (v - b)^2 = r^2 \text{ and } (u - c)^2 + (v - d)^2 = s^2.$$

After expanding, both equations have the form $u^2 + v^2 + \text{something} = 0$. Setting the something's equal gives an equation of the form $tu + t'v + t'' = 0$, where $t, t', t'' \in F$. Coupling this with the equation of one of the circles returns us to the situation of the second paragraph. $\bullet$

Theorem R.7. *A complex number z is constructible if and only if there is a tower of fields*

$$\mathbb{Q} = K_0 \subset K_1 \subset \cdots \subset K_n,$$

where $z \in K_n$ and $[K_{j+1} : K_j] \leq 2$ for all j.

Proof. If z is constructible, there is a sequence of points $1, -1, z_1, \ldots, z_n = z$ with each z_j obtainable from $\{1, -1, z_1, \ldots, z_{j-1}\}$; since i is constructible, we may assume that $z_1 = i$. Define

$$K_j = \mathbb{Q}(z_1, \ldots, z_j).$$

Given $u = z_{j+1}$, there are points $E, F, G, H \in K_j$ with one of the following:

$$\begin{aligned} u &\in L(E, F) \cap L(G, H); \\ u &\in L(E, F) \cap C(G; H); \\ u &\in C(E; F) \cap C(G; H). \end{aligned}$$

We may assume, by induction on $j \geq 1$, that K_j is closed under complex conjugation, so that Lemma R.6 applies to show that $[K_{j+1} : K_j] \leq 2$. Finally, note that K_{j+1} is also closed under complex conjugation, for if z_{j+1} is a root of a quadratic $f(x) \in K_j[x]$, then $\overline{z}_{j+1}$ is the other root of $f(x)$.

To prove the converse, it suffices to prove that if $[B : F] = 2$, where $F \subset K$, then B/F is a pure extension of type 2, say, $B = F(\beta)$, where $\beta \in L(P, Q) \cap C(R; S)$ for $P, Q, R, S \in F$; it will then follow that $B \subset K$. Since $[B : F] = 2$, there is α with $B = F(\alpha)$, where α is a root of some irreducible quadratic $x^2 + bx + c \in F[x]$. If we define $\beta = \sqrt{b^2 - 4c}$, then $B = F(\beta)$ displays B/F as a pure extension of type 2. To see that β can be realized as a point in the intersection of a line and a circle, we use the construction in Theorem R.3(v). Let the line L be the vertical line through $A = (1, 0)$ and let the circle have center $Q = (\frac{1}{2}(1 + \beta^2), 0)$ and radius $\frac{1}{2}(1 + \beta^2)$. •

Corollary R.8. *If a complex number z is constructible, then $[\mathbb{Q}(z) : \mathbb{Q}]$ is a power of 2.*

Proof. This follows from the theorem and Lemma 49. •

Remark. The converse of this corollary is false. In Example 36, we saw that $p(x) = x^4 - 4x + 2$ is an irreducible polynomial over $\mathbb{Q}$ whose Galois group $\text{Gal}(E/\mathbb{Q})$ is S_4, where $E/\mathbb{Q}$ is a splitting field of $p(x)$. Were every root of $p(x)$ constructible, then every element of E would be constructible, for all constructible numbers form a subfield of $\mathbb{C}$, by Theorem R.3. If H is a Sylow 2-subgroup of $G \cong S_4$, however, then $[G : H] = 3$; the intermediate field E^H thus has degree $[E^H : \mathbb{Q}] = [G : H] = 3$, and so none of its elements are constructible, by Corollary R.8. This contradiction shows that some root of $p(x)$ is not constructible, even though every root has degree 4 over $\mathbb{Q}$.

It is now a simple matter to dispose of some famous problems.

(1) *It is impossible to "square the circle."*

The problem is to construct, with ruler and compass, a square whose area is equal to the area of a circle of radius 1; in other words, one asks whether $\sqrt{\pi}$ is constructible. But it is a classical result, proved by F. Lindemann in 1882, that π, hence $\sqrt{\pi}$, is transcendental over $\mathbb{Q}$ (see [Hadlock, p. 47]), and so it does not lie in any finite extension of $\mathbb{Q}$, let alone one of degree a power of 2.

(2) It is impossible to "duplicate the cube."

The problem is to construct a cube whose volume is 2; in other words, is the real cube root of 2, call it α, constructible? Now $x^3 - 2$ is irreducible over $\mathbb{Q}$, by Eisenstein, and so $[\mathbb{Q}(\alpha) : \mathbb{Q}] = 3$, which is not a power of 2. Corollary R.8 gives the result. This result was first proved by P. L. Wantzel in 1837.

(3) It is impossible to trisect an arbitrary angle.

An angle θ is given by two intersecting lines; it is no loss in generality to assume the lines intersect at the origin and that one line is the x-axis. If we could draw the angle trisector, then the point $(\cos \theta/3, \sin \theta/3)$, which is the intersection of the trisector and the unit circle, would be constructible; hence $\cos \theta/3$ would also be constructible, by Lemma R.1.

We will now show that $60°$ cannot be trisected. Computing the real part of $e^{3i\theta} = (\cos \theta + i \sin \theta)^3$ gives the trigonometric identity:

$$\cos 3\theta = 4\cos^3 \theta - 3\cos \theta.$$

Defining $u = 2 \cos \theta$ and $\theta = 20°$, we arrive at the equation

$$u^3 - 3u - 1 = 0.$$

It is easy to see that this cubic is irreducible (it has no rational root, by Exercise 63), and so $[\mathbb{Q}(u) : \mathbb{Q}] = 3$. Corollary R.8 shows that u is not constructible. This result was also proved by P. L. Wantzel in 1837.

(4) Regular p-gons.

Galois theory will be used in discussing this problem.

Theorem R.9 (Gauss). *If p is an odd prime, then a regular p-gon is constructible if and only if $p = 2^{2^t} + 1$ for some $t \geq 0$.*

Proof. This is again a question of constructibility of a point on the unit circle, namely, $z = e^{2\pi i/p}$. Now the irreducible polynomial of z over $\mathbb{Q}$ is the cyclotomic polynomial $\Phi_p(x)$ of degree $p - 1$ (Corollary 41).

Assume z is constructible. By Corollary R.8, $p - 1 = 2^s$ for some s. We claim that s itself is a power of 2. Otherwise, there is an odd number $k > 1$ with $s = km$. But $x^k + 1$ factors over $\mathbb{Z}$ (because -1 is a root); setting $x = 2^m$ thus gives a forbidden factorization of p.

Conversely, assume $p = 2^{2^t} + 1$ is prime. Since z is a primitive pth root of unity, $\mathbb{Q}(z)$ is the splitting field of $\Phi_p(x)$ over $\mathbb{Q}$. Hence $\mathrm{Gal}(\mathbb{Q}(z)/\mathbb{Q})$ has order 2^{2^t}, and so the Galois group is a 2-group. But a 2-group has a normal series in which each factor group has order 2 (this follows easily from Theorem G.23); by the fundamental theorem of Galois theory, there is a tower of fields $\mathbb{Q} = K_0 \subset K_1 \subset \cdots \subset K_m = \mathbb{Q}(z)$ with $[K_{i+1} : K_i] = 2$ for all i, that is, z is constructible, by Theorem R.7.   •

Remark. Primes of the form $2^{2^t} + 1$ are called **Fermat primes**. The values $0 \leq t \leq 4$ do give primes (they are 3, 5, 17, 257, 65,537), the next few values of t do not give primes, and it is unknown whether any other Fermat primes exist.

Gauss actually gave a geometric construction of the regular 17-gon.

Corollary R.10. *It is impossible to construct a regular 7-gon, a regular 11-gon, or a regular 13-gon.*

Proof. 7, 11, and 13 are not Fermat primes.   •

The following result is known (see [Hadlock, p. 106]):

Theorem R.11. *A regular n-gon is constructible if and only if n is a product of a power of 2 and distinct Fermat primes.*

It follows that regular 9-gons and regular 14-gons are not constructible; on the other hand, a regular 15-gon is constructible. It is possible that there are only finitely many constructible regular n-gons with n is odd, for there may be only finitely many Fermat primes.

Appendix D
Old-fashioned Galois Theory

Gimme that old-time Galois theory;

If it's good enough for Galois, then it's good enough for me!

I am a creature of the twentieth century; algebraic systems and their auto-morphism groups are part of my mother's milk. When writing the defini-tion of Galois group for this text, I asked myself an obvious question: how did such thoughts occur to Galois in the late 1820's? The answer, of course, is that he did not think in such terms; for its first century, 1830–1930, the Galois group was a group of permutations. In the late 1920's, E. Artin, de-veloping ideas of E. Noether going back at least to Dedekind, recognized that it is both more elegant and more fruitful to describe Galois groups in terms of field automorphisms (Artin's version is isomorphic to the original version). In 1930, van der Waerden incorporated much of Artin's view-point into his influential text "Moderne Algebra," and a decade later Artin published his own lectures. So successful have Artin's ideas proved to be that they have virtually eclipsed earlier expositions. But we have lost the inevitability of the definition; group theory is imposed on the study of poly-nomials rather than arising naturally from it. This appendix is an attempt to remedy this pedagogical problem by telling the story of what happened in the beginning. The reader interested in a more thorough account may read [Edwards] or [Tignol].

We use modern notation and terms even though they were unknown in the late eighteenth century. In particular, F shall denote a subfield of the complex numbers. Permutations arise simultaneously with the question of finding the roots of a polynomial. If

$$f(x) = \prod_{i=1}^{n} (x - \alpha_i) = x^n + b_{n-1}x^{n-1} + \cdots + b_1 x + b_0,$$

then one sees easily that b_{n-j} is, to sign, the sum of all products of j roots α_i:

$$b_{n-j} = (-1)^j \sum_{1 \le i_1 < i_2 < \cdots < i_j \le n} \alpha_{i_1} \alpha_{i_2} \cdots \alpha_{i_j}.$$

Thus

$$b_{n-1} = -\sum_i \alpha_i = -(\alpha_1 + \cdots + \alpha_n)$$

$$b_{n-2} = \sum_{i<j} \alpha_i \alpha_j$$

$$b_{n-3} = -\sum_{i<j<k} \alpha_i \alpha_j \alpha_k$$

$$\vdots$$

$$b_0 = (-1)^n \alpha_1 \alpha_2 \cdots \alpha_n.$$

Since the coefficients b_{n-j} are unchanged if the roots are re-indexed, it is clear that they are symmetric functions of the roots in the following sense.

Definition. A polynomial $g(x_1, \ldots, x_n) \in F[x_1, \ldots, x_n]$ is **symmetric** if

$$g(x_{\sigma 1}, \ldots, x_{\sigma n}) = g(x_1, \ldots, x_n)$$

for every $\sigma \in S_n$.

Each of the polynomials

$$e_j(x_1, \ldots, x_n) = \sum_{1 \le i_1 < i_2 < \cdots < i_j \le n} x_{i_1} x_{i_2} \cdots x_{i_j}$$

is symmetric; one calls $e_1, \ldots, e_n$ the **elementary symmetric functions**. Note that $e_j(\alpha_1, \ldots, \alpha_n) = (-1)^j b_{n-j}$.

The following result was well known in the late 1700's. For a proof, see [Hadlock, p. 42].

Theorem H.1 (Fundamental Theorem of Symmetric Functions). *If* $g(x_1, \ldots, x_n) \in F[x_1, \ldots, x_n]$ *is symmetric, then there exists*

$$h(x_1, \ldots, x_n) \in F[x_1, \ldots, x_n],$$

not necessarily symmetric, with

$$g(x_1, \ldots, x_n) = h(e_1, \ldots, e_n).$$

In 1770, Waring published an algorithm for finding h. For example,

$$
\begin{aligned}
x_1^2 + x_2^2 + x_3^2 &= (x_1 + x_2 + x_3)^2 - 2(x_1 x_2 + x_1 x_3 + x_2 x_3) \\
&= e_1^2 - 2e_2.
\end{aligned}
$$

Corollary H.2. *Let* $f(x) = x^n + b_{n-1}x^{n-1} + \cdots + b_1 x + b_0 \in F[x]$ *have (complex) roots* $\alpha_1, \ldots, \alpha_n$; *if* $g(x_1, \ldots, x_n) \in F[x_1, \ldots, x_n]$ *is symmetric, then*

$$g(\alpha_1, \ldots, \alpha_n) \in F.$$

Proof. There is $h(x_1, \ldots, x_n) \in F[x_1, \ldots, x_n]$, by the fundamental theorem, with $g(x_1, \ldots, x_n) = h(e_1, \ldots, e_n)$. Specializing $(x_1, \ldots, x_n)$ to $(\alpha_1, \ldots, \alpha_n)$ gives $g(\alpha_1, \ldots, \alpha_n) = h(-b_{n-1}, \ldots, \pm b_0) \in F$. •

The classical formulas for the roots of cubics and quartics, discovered more than two centuries earlier, were also well known. Recall that the roots of $f(x) = x^3 + qx + r$ are:

$$\alpha_1 = y + z; \quad \alpha_2 = \omega y + \omega^2 z; \quad \alpha_3 = \omega^2 y + \omega z;$$

here, $y^3 = \frac{1}{2}(-r + \sqrt{R})$ [where $R = r^2 + 4q^3/27$], $z = -q/3y$, and ω is a primitive cube root of unity. In 1770, Lagrange and Vandermonde, independently, sought to find the basic principles underlying the known formulas. They expressed the radicals in terms of the roots α_i:

$$3z = \alpha_1 + \omega\alpha_2 + \omega^2\alpha_3;$$
$$3y = \alpha_1 + \omega^2\alpha_2 + \omega\alpha_3.$$

For given $\alpha_1, \alpha_2, \alpha_3$ and (not necessarily primitive) cube root of unity ω, let us denote

$$\psi(\omega) = (\alpha_1 + \omega\alpha_2 + \omega^2\alpha_3)^3;$$

then

$$(3z)^3 = \psi(\omega) \quad \text{and} \quad (3y)^3 = \psi(\omega^2)$$

and, for $i = 1, 2, 3$,

$$\alpha_i = \frac{1}{3}\left(\sqrt[3]{\psi(\omega)} + \sqrt[3]{\psi(\omega^2)} \right).$$

How can one determine the two numbers $\psi(\omega)$ and $\psi(\omega^2)$? Regard the roots $\alpha_1, \alpha_2, \alpha_3$ as indeterminates and define:

$$\varphi_1(x_1, x_2, x_3) = x_1 + \omega x_2 + \omega^2 x_3;$$
$$\varphi_2(x_1, x_2, x_3) = x_1 + \omega^2 x_2 + \omega x_3.$$

Neither φ_1 nor φ_2 is symmetric. Now the transposition (23) interchanges φ_1 and φ_2, because (23) sends $x_1 \mapsto x_1$, $x_2 \mapsto x_3$ and $x_3 \mapsto x_2$. The 3-cycle (132) fixes both φ_1^3 and φ_2^3; for example, (132) sends φ_1^3 into

$$(x_3 + \omega x_1 + \omega^2 x_2)^3 = [\omega(\omega^2 x_3 + x_1 + \omega x_2)]^3 = \varphi_1^3;$$

this is one reason for cubing φ_1 and φ_2. It follows that $\varphi_1^3 + \varphi_2^3$ and $\varphi_1^3\varphi_2^3$ are symmetric functions [each is invariant under (23) and (132), and these two permutations generate the symmetric group S_3]. The algorithm for the fundamental theorem of symmetric functions expresses $\varphi_1^3 + \varphi_2^3$ and $\varphi_1^3\varphi_2^3$ in terms of elementary symmetric functions. Since $\varphi_1(\alpha_1, \alpha_2, \alpha_3)^3 = \psi(\omega)$

and $\varphi_2(\alpha_1, \alpha_2, \alpha_3)^3 = \psi(\omega^2)$, the corollary of the fundamental theorem expresses $b_1 = \psi(\omega) + \psi(\omega^2)$ and $b_0 = \psi(\omega)\psi(\omega^2)$ in terms of the coefficients q and r of $f(x)$. We have seen that once we know $\psi(\omega)$ and $\psi(\omega^2)$, we can find the roots $\alpha_1, \alpha_2, \alpha_3$ of $f(x)$. But

$$x^2 - b_1 x + b_0 = (x - \psi(\omega))(x - \psi(\omega^2)),$$

and so $\psi(\omega)$ and $\psi(\omega^2)$ can be found by the quadratic formula. (There are four more polynomials obtained from $\varphi_1(x_1, x_2, x_3)$ by permuting variables: $\omega\varphi_1; \omega^2\varphi_1; \omega\varphi_2; \omega^2\varphi_2$. These are the other cube roots of $\psi(\omega)$ and $\psi(\omega^2)$; using them replaces $3y$ by $3\omega y$ and $3z$ by $3\omega^2 z$, for example, and this merely reindexes the α's.)

Both Lagrange and Vandermonde did a similar analysis of the quartic. If the roots are $\alpha_1, \alpha_2, \alpha_3, \alpha_4$, then they defined

$$\varphi_1(x_1, x_2, x_3, x_4) = x_1 + i x_2 + i^2 x_3 + i^3 x_4$$

where $i^2 = -1$ (i.e., i is a 4th root of unity), and they showed that φ_1^4 plays a decisive role in obtaining the classical formula.

Lagrange generalized this analysis to polynomials

$$f(x) = x^n + b_{n-1}x^{n-1} + \cdots + b_0$$

of degree n. If ω is an nth root of unity (not necessarily primitive) and if the roots of $f(x)$ are $\alpha_1, \ldots, \alpha_n$, define numbers

$$
\begin{aligned}
\varphi_1(\omega) &= \alpha_1 + \alpha_2\omega + \alpha_3\omega^2 + \cdots + \alpha_n\omega^{n-1}, \\
\varphi_2(\omega) &= \alpha_2 + \alpha_3\omega + \alpha_4\omega^2 + \cdots + \alpha_1\omega^{n-1}, \\
&\;\;\vdots \\
\varphi_n(\omega) &= \alpha_n + \alpha_1\omega + \alpha_2\omega^2 + \cdots + \alpha_{n-1}\omega^{n-1}.
\end{aligned}
$$

Notice that

$$\varphi_i(\omega) = \omega^{n-i+1}\varphi_1(\omega),$$

and that

$$\varphi_1(1) = \alpha_1 + \alpha_2 + \cdots + \alpha_n = -b_{n-1}.$$

By analogy with the analysis of the cubic, Lagrange defined

$$\psi(\omega) = [\varphi_1(\omega)]^n.$$

Note that $[\varphi_i(\omega)]^n = [\varphi_1(\omega)]^n$ for $1 \leq i \leq n$. Of course, $\psi(1) = [\varphi_1(1)]^n = [-b_{n-1}]^n$ is known.

Lemma H.3. *If $f(x)$ has degree n, then the roots $\alpha_1, \ldots, \alpha_n$ of $f(x)$ are determined by the $n - 1$ numbers*

$$\psi(\omega), \psi(\omega^2), \ldots, \psi(\omega^{n-1}),$$

where ω is a primitive nth root of unity.

Proof. Consider

$$\sum_{j=0}^{n-1} \varphi_1(\omega^j) = \sum_{j=0}^{n-1} (\alpha_1 + \alpha_2\omega^j + \alpha_3\omega^{2j} + \cdots + \alpha_n\omega^{(n-1)j})$$

$$= n\alpha_1 + \alpha_2 \sum \omega^j + \alpha_3 \sum \omega^{2j} + \cdots + \alpha_n \sum \omega^{(n-1)j}.$$

For each fixed k with $1 \le k \le n - 1$, the geometric series $\sum_{j=0}^{n-1} \omega^{kj}$ sums to $\dfrac{1 - (\omega^k)^n}{1 - \omega^k} = 0$, because $\omega^{kn} = 1$ and $\omega^k \ne 1$. It follows that

$$n\alpha_1 = \sum_{j=0}^{n-1} \varphi_1(\omega^j).$$

Similarly, $n\alpha_i = \sum_j \varphi_i(\omega^j)$. But $\varphi_i(\omega) = \omega^{n-i+1}\varphi_1(\omega)$, so that

$$n\alpha_i = \sum_{j=0}^{n-1} \varphi_i(\omega^j) = \omega^{n-i+1} \sum_{j=0}^{n-1} \varphi_1(\omega^j).$$

Therefore, the roots $\alpha_1, \ldots, \alpha_n$ are determined by $\varphi_1(1), \varphi_1(\omega), \varphi_1(\omega^2), \ldots, \varphi_1(\omega^{n-1})$. But $\varphi_1(1) = -b_{n-1}$, so that $\alpha_1, \ldots, \alpha_n$ are determined by the $n - 1$ numbers $\varphi_1(\omega), \varphi_1(\omega^2), \ldots, \varphi_1(\omega^{n-1})$. Finally, $\varphi_1(\omega) = \sqrt[n]{\psi(\omega)}$, and so the roots are determined by the $n - 1$ numbers $\psi(\omega), \psi(\omega^2), \ldots, \psi(\omega^{n-1})$. •

The last lemma, essentially due to Bézout (1765), says that the n roots of $f(x)$, a polynomial of degree n, can be found in terms of $n - 1$ numbers $\psi(\omega), \ldots, \psi(\omega^{n-1})$; that is, there is a polynomial of degree $n - 1$, namely,

$$\rho(x) = \prod_{j=1}^{n-1} [x - \psi(\omega^j)],$$

whose roots determine the roots of $f(x)$. Does this not give the inductive step for finding the roots of a polynomial of arbitrary degree n? The answer, unfortunately, is negative because we do not know the coefficients of

$\rho(x)$.[19] At the very least, we need these coefficients to lie in F; and it is precisely this that introduces groups into the theory! Lagrange's idea was to replace $\rho(x)$ by a more manageable polynomial in $F[x]$.

The number $\psi(\omega) = (\alpha_1 + \alpha_2\omega + \cdots + \alpha_n\omega^{n-1})^n$ is not symmetric in the α; let us try to force it to be. If $g(x_1, \ldots, x_n) \in F[x_1, \ldots, x_n]$ and $\sigma \in S_n$, define a polynomial σg by

$$\sigma g(x_1, \ldots, x_n) = g(x_{\sigma 1}, \ldots, x_{\sigma n});$$

just permute the indeterminates as σ prescribes. Now consider the "symmetrized" polynomial of $n + 1$ variables

$$g^*(x, x_1, \ldots, x_n) = \prod_{\sigma \in S_n} [x - \sigma g(x_1, \ldots, x_n)];$$

its coefficients have the form

$$e(\sigma_1 g(x_1, \ldots, x_n), \ldots, \sigma_{n!} g(x_1, \ldots, x_n)),$$

where e is an elementary symmetric function of the $n!$ terms $\sigma g(x_1, \ldots, x_n)$ and the permutations in S_n are listed $\sigma_1, \sigma_2, \ldots, \sigma_{n!}$. If τ is any permutation in S_n, then

$$e(\sigma_1 g(x_{\tau 1}, \ldots, x_{\tau n}), \ldots, \sigma_{n!} g(x_{\tau 1}, \ldots, x_{\tau n}))$$
$$= e(\sigma_1 \tau g(x_1, \ldots, x_n), \ldots, \sigma_{n!} \tau g(x_1, \ldots, x_n)).$$

As σ_i varies over all of S_n, so does $\sigma_i \tau$. Permuting the x_i by τ thus permutes the coordinates in the argument of e; as e is symmetric, it follows that the coefficients of $g^*(x, x_1, \ldots, x_n)$ are symmetric in the x_i. Specializing $(x_1, \ldots, x_n)$ to $(\alpha_1, \ldots, \alpha_n)$ thus yields a polynomial $\tilde{g}(x) \in F[x]$, by Corollary H.2. Although the degree of $\tilde{g}(x)$ is large (it is $n!$), it does have one important property: any one of its roots $g(\alpha_1, \ldots, \alpha_n)$ determines all of the others because we know $g(x_1, \ldots, x_n)$ and

$$\sigma g(\alpha_1, \ldots, \alpha_n) = g(\alpha_{\sigma 1}, \ldots, \alpha_{\sigma n})$$

for $\sigma \in S_n$.

In particular, regard $\psi(\omega) = (\alpha_1 + \alpha_2\omega + \cdots + \alpha_n\omega^{n-1})^n$ as a function of n indeterminates. Then

$$\psi^*(x, x_1, \ldots, x_n) = \prod_\sigma [x - \sigma\psi(x_1, \ldots, x_n)]$$

---

[19]We gave an argument above that these coefficients are known when $n = 3$.

is a polynomial in x with coefficients in $F(\omega)(x_1, \ldots, x_n)$, the field of fractions of $F(\omega)[x_1, \ldots, x_n]$; specializing $(x_1, \ldots, x_n)$ to $(\alpha_1, \ldots, \alpha_n)$ gives a polynomial $\tilde{\psi}(x)$ in $F(\omega)[x]$.

One of the roots of $\tilde{\psi}(x)$ is $\psi(\omega)$. Assume now that n is prime. If $1 \leq j \leq n-1$, then ω^j is a primitive nth root of unity; hence $\omega^j, \omega^{2j}, \ldots, \omega^{(n-1)j}$ is a permutation, say σ, of $\omega, \omega^2, \ldots, \omega^{n-1}$, and so

$$\psi(\omega^j) = \sigma\psi(\omega).$$

It follows that $\psi(\omega), \psi(\omega^2), \ldots, \psi(\omega^{n-1})$ are roots of $\tilde{\psi}(x)$. (This same argument applies to any n if one chooses j relatively prime to n.)

If $g(x_1, \ldots, x_n) \in F[x_1, \ldots, x_n]$, then $g^*(x, x_1, \ldots, x_n)$ can be simplified by eliminating repetitions: if $\sigma_i g = \sigma_k g$, throw away one of them.

Definition. A polynomial $g(x_1, \ldots, x_n)$ is r-*valued*,[20] where $1 \leq r \leq n!$, if there are exactly r distinct polynomials of the form σg for $\sigma \in S_n$.

Thus, 1-valued functions are symmetric functions, while the function

$$\Delta(x_1, \ldots, x_n) = \prod_{i<j}(x_i - x_j)$$

is always 2-valued. Note that $\psi(x_1, x_2, x_3) = (x_1 + x_2\omega + x_3\omega^2)^3$ is 2-valued, $g(x_1, x_2, x_3) = x_1$ is 3-valued, and $h(x_1, x_2, x_3) = x_1x_2 - x_2x_3$ is 6-valued.

Plainly, if ψ is r-valued, then $\tilde{\psi}(x)$ should be replaced by its factor of degree r, call it $\lambda(x)$, which is obtained from $\tilde{\psi}(x)$ by discarding repeated factors; $\lambda(x)$ is called the **Lagrange resolvent** of $f(x)$; this is Lagrange's replacement for the polynomial $\rho(x)$ of degree $n-1$. How can we compute its degree r?

Definition. If $g(x_1, \ldots, x_n) \in F[x_1, \ldots, x_n]$, then

$$G(g) = \{\sigma \in S_n : \sigma g = g\}.$$

---

[20]This is the standard terminology occurring in all the older references. Do not confuse it with modern usage which, for example, calls the relation (not a function) $f(x) = \pm\sqrt{x}$ a 2-valued function.

Lagrange claimed (but his proof is incomplete) that [21]

$$r = n!/|G(g)|.$$

In particular, a polynomial $g(x_1, \ldots, x_n)$ is an $n!$-valued function if $G(g) = \{1\}$.

There are two ways of regarding a permutation of n letters. The first way is as a list of length n having no repetitions; the second way is as a bijection. The latter version invites composition: one can multiply two permutations to get a third one. It seems likely that Lagrange was not aware that $G(g)$ is a subgroup of S_n, for he was viewing permutations as lists.

Lagrange did prove a remarkable theorem showing the importance of $G(g)$.

Theorem H.4 (Lagrange's Rational Function Theorem).
If $g, h \in F[x_1, \ldots, x_n]$, then $G(h) \subset G(g)$ if and only if g is a rational function of h; that is, there is a rational function $\theta(x)$ with $g = \theta(h)$, where the coefficients of $\theta(x)$ involve F and are symmetric functions of $x_1, \ldots, x_n$.

Corollary H.5. *If $g, h \in F[x_1, \ldots, x_n]$, then $G(g) = G(h)$ if and only if each of g and h is a rational function of the other.*

Corollary H.6. *If $h \in F[x_1, \ldots, x_n]$ is an $n!$-valued function, then every $g \in F[x_1, \ldots, x_n]$ is a rational function of h.*

Corollary H.7. *If $h \in F[x_1, \ldots, x_n]$ is an $n!$-valued function, then each x_i is a rational function of h.*

Corollary H.8 (Theorem of Primitive Element). *If $\alpha_1, \ldots, \alpha_n$ are the roots of $f(x) \in F[x]$, then there exists η with $F(\alpha_1, \ldots, \alpha_n) = F(\eta)$.*
Moreover, there exist rational functions $\theta_i(x) \in F(x)$ with $\alpha_i = \theta_i(\eta)$ for all $i = 1, \ldots, n$.

---

[21]Here is a modern proof. The group S_n acts on $F[x_1, \ldots, x_n]$ by permuting the variables; $G(g)$ is the stabilizer of g; r is the size of the orbit of g. Theorem G.10 gives

$$r = [S_n : G(g)] = n!/|G(g)|.$$

This claim is also the reason Lagrange's theorem in group theory is so-called.

Proof. Let $h(x_1, \ldots, x_n) \in F[x_1, \ldots, x_n]$ be an $n!$-valued function; for each i, define $g_i(x_1, \ldots, x_n) \in F[x_1, \ldots, x_n]$ by $g_i(x_1, \ldots, x_n) = x_i$. By Corollary H.7, there exist rational functions $\theta_i(x) \in F(x)$ with

$$x_i = g(x_1, \ldots, x_n) = \theta_i(h(x_1, \ldots, x_n)).$$

Define $\eta = h(\alpha_1, \ldots, \alpha_n)$. •

Let us summarize this 1770 work of Lagrange. A polynomial $f(x) \in F[x]$ of degree n determines a polynomial ψ of n variables. This polynomial determines a subgroup $G(\psi)$ of S_n; "symmetrizing" ψ gives a polynomial $\widetilde{\psi}(x) \in F(\omega)[x]$ whose roots, when n is prime, suffice to find the roots of $f(x)$. Discarding repeated roots of $\widetilde{\psi}(x)$ leaves the Lagrange resolvent $\lambda(x) \in F(\omega)[x]$, a polynomial of degree r, and knowledge of one root of $\lambda(x)$ determines the other roots.

Lagrange had hoped that his procedure might solve the general polynomial of degree n. On the other hand, his analysis of the quintic led him to an intractible sextic, with no obvious way to find a root, and this discouraged him.

There was progress in the sixty years from Lagrange to Galois. In 1803, Gauss analyzed roots of unity and cyclotomic polynomials (one consequence is the determination of those regular polygons constructible by ruler and compass). Ruffini (1799) and Abel (1824) essentially proved the insolvability of the general quintic (neither proof is correct in all details, but Abel's proof was accepted and Ruffini's was not). In 1829, Abel proved that certain polynomials $f(x)$ are always solvable by radicals: if $\alpha_1, \ldots, \alpha_n$ are the roots of $f(x)$, if there are rational functions $\theta_i(x)$ with $\alpha_i = \theta_i(\alpha_1)$ for all $i = 1, \ldots, n$, and if

$$\theta_i(\theta_j(\alpha_1)) = \theta_j(\theta_i(\alpha_1))$$

for all i, j (in modern language, the Galois group is abelian; this result is the etymology of the adjective). (See [Tignol; p. 316] for more discussion.)

Although group theory did not exist before Galois, there were some results which today can be seen as group theoretic. Ruffini showed that there are no r-valued functions of 5 variables for $r = 3, 4$, and 8; that is, S_5 has no subgroups of index 3, 4, or 8 (see Lemma G.38). Abbati (1803) proved that $|G(g)|$ does, indeed, divide $n!$, so that Lagrange's assertion about the degree r is correct. Thus, Abbati proved "Lagrange's Theorem" (Theorem G.3) for subgroups of S_n; the full theorem was probably first proved by Galois. Abbati also proved: A_n is the only subgroup of S_n

having index 2; If $n \geq 5$, then S_n has no subgroups of index 3 or 4. Cauchy (1815) established the calculus of permutations, e.g., decomposition into disjoint cycles; he proved that, for n prime, S_n has no subgroups of index r with $2 < r < n$.

Galois knew that some polynomials are solvable by radicals and some are not; it was reasonable that it depends on the roots. The Lagrange resolvent $\lambda(x)$ is not sensitive to this. Indeed, it seems that Lagrange was seeking a formula for the roots of the *general polynomial* $x^n + b_{n-1}x^{n-1} + \cdots + b_0$: the roots of any particular polynomial $f(x)$ of degree n would be obtained from the "master formula" by substituting the specific coefficients of $f(x)$. (The classical formulas for polynomials of degree ≤ 4 are of this form.) If $f(x) \in F[x]$ has roots $\alpha_1, \ldots, \alpha_n$, Lagrange first regarded $\alpha_1, \ldots, \alpha_n$ as indeterminates, then he formed $\psi(x_1, \ldots, x_n) = (x_1 + \omega x_2 + \omega^2 x_3 + \cdots + \omega^{n-1}x_n)^n$, symmetrized to obtain

$$\psi^*(x, x_1, \ldots, x_n) = \prod_{\sigma \in S_n} [x - \sigma\psi(x_1, \ldots, x_n)],$$

defined $\lambda(x)$ to be the factor of ψ^* of degree r (in x) which is the product over all distinct polynomials $\sigma\psi$, and finally specialized $(x_1, \ldots, x_n)$ back to $(\alpha_1, \ldots, \alpha_n)$. But even if $\sigma\psi(x_1, \ldots, x_n)$ and $\tau\psi(x_1, \ldots, x_n)$ are distinct polynomials, the numbers $\sigma\psi(\alpha_1, \ldots, \alpha_n) = \tau\psi(\alpha_1, \ldots, \alpha_n)$ may be equal. As a polynomial over $F(\omega)(x_1, \ldots, x_n)$, the Lagrange resolvent $\lambda(x) = \lambda(x; x_1, \ldots, x_n)$ has distinct roots; $\lambda(x) = \lambda(x; \alpha_1, \ldots, \alpha_n)$, as a polynomial over F, may have repeated roots. One can discard these extra roots but, unfortunately,

$$\{\sigma \in S_n : (\sigma\psi)(\alpha_1, \ldots, \alpha_n) = \psi(\alpha_1, \ldots, \alpha_n)\}$$

may not be a subgroup of S_n and this prevents the generalization of Lagrange's Rational Function Theorem from being true.

Galois jettisoned $\psi(x_1, \ldots, x_n)$ which, after all, works best when the degree n is prime; he replaced it by an $n!$-valued function $V(x_1, \ldots, x_n)$ with an added property: all $(\sigma V)(\alpha_1, \ldots, \alpha_n)$ are distinct (of course, this forces all the α_i to be distinct; this minor point is easily handled by Exercise 44). Let us call (after Edwards) such a function V a **Galois resolvent**[22] of $f(x)$. Galois knew that such resolvents exist (Lagrange had proven it); indeed, there are such of the form $V(x_1, \ldots, x_n) = c_1x_1 + \cdots + c_nx_n$,

---

[22] Actually, I would prefer that the polynomial $\gamma(x)$ below be called the Galois resolvent, for it is analogous to $\lambda(x)$ whereas V is analogous to ψ.

for suitable $c_1, \ldots, c_n \in F$. Denote $V(\alpha_1, \ldots, \alpha_n)$ by v_1. Since V is $n!$-valued, there are rational functions $\theta_1(x), \ldots, \theta_n(x)$ in $F(x)$ with $\alpha_i = \theta_i(v_1)$ for all i.

The next step ought to be the symmetrization of V: define

$$V^*(x; x_1, \ldots, x_n) = \prod_{\sigma \in S_n} [x - \sigma V(x_1, \ldots, x_n)],$$

and then choose a factor of V^* by discarding repeated roots. Galois did this indirectly. Let $\gamma(x)$ be the irreducible polynomial of v_1 over F, and let $v_1, \ldots, v_m$ be the roots of $\gamma(x)$.

Recall Exercise 55: Let $f(x), g(x) \in F[x]$. Then $(f, g) \neq 1$ if and only if there is a field E containing both F and a common root of $f(x)$ and $g(x)$. It follows that if $p(x)$ is irreducible, then $p(x)$ divides $h(x)$. Therefore, $\gamma(x)$ divides $\widetilde{V}(x) = V^*(x; \alpha_1, \ldots, \alpha_n)$, and so each root v_j of $\gamma(x)$ has the form $\sigma V(\alpha_1, \ldots, \alpha_n)$ for some permutation $\sigma \in S_n$. But Galois wanted a more explicit description of σ. Here is an easy generalization of Exercise 50: Let $p(x) \in F[x]$ be an irreducible polynomial and let $\Phi(x) \in F(x)$ be a rational function; if $\Phi(v) = 0$ for some root v of $p(x)$, then $\Phi(v') = 0$ for every root v' of $p(x)$.

Theorem H.9. *Let $f(x) \in F[x]$ have distinct roots $\alpha_1, \ldots, \alpha_n$, and let $v_1, \ldots, v_m$ be as above; let $\alpha_i = \theta_i(v_1)$, where $\theta_i(x) \in F(x)$ for all i. Then for each $j = 1, \ldots, m$, the function*

$$\sigma_j : \alpha_i = \theta_i(v_1) \mapsto \theta_i(v_j), \qquad i = 1, \ldots, n,$$

is a permutation of the roots $\alpha_1, \ldots, \alpha_n$.

Proof. Define $\Phi(x) \in F(x)$ by $\Phi(x) = f(\theta_i(x))$. Now

$$\Phi(v_1) = f(\theta_i(v_1)) = f(\alpha_i) = 0;$$

since $\gamma(x)$ is irreducible, the generalized Exercise 50 shows that $0 = \Phi(v_j) = f(\theta_i(v_j))$; that is, $\theta_i(v_j)$ is a root of $f(x)$, hence is one of the α's. To see that σ_j is a permutation, it suffices to prove it is an injection. Suppose that $\theta_i(v_j) = \theta_k(v_j)$. Now $\Phi(x) = \theta_i(x) - \theta_k(x)$ is a rational function with $\Phi(v_j) = 0$; it follows that $0 = \Phi(v_1) = \theta_i(v_1) - \theta_k(v_1) = \alpha_i - \alpha_k$. Since all the roots of $f(x)$ are distinct, $i = k$, as desired. •

Galois defined the Galois group of $f(x)$ as

$$\text{Gal}(f) = \{\text{all } \sigma_j : \alpha_i = \theta_i(v_1) \mapsto \theta_i(v_j)\}.$$

This is the beginning of Galois's 1831 paper in which he characterizes polynomials solvable by radicals as those having a solvable Galois group. (For a proof that this definition is equivalent to the modern one in terms of automorphisms, see [Tignol, p. 329].)

Subtle group theoretic clues were in the air, but only Galois recognized their significance; developing them, he invented group theory and solved the mystery of the roots of polynomials. This is even more impressive when we realize that this is no less than the birth of modern algebra.

References

[1] E. Artin, *Galois Theory (second edition)*, Notre Dame, 1955.

[2] G. Birkhoff and S. Mac Lane, *A Survey of Modern Algebra, (fourth edition)*, Macmillan, 1977.

[3] W. S. Burnside and A. W. Panton, *The Theory of Equations, vol. II*, Longmans, Green, 1899.

[4] S. Chase, D. Harrison, and A. Rosenberg, *Galois Theory and Cohomology of Commutative Rings*, Mem. Amer. Math. Soc., 1965.

[5] E. Dehn, *Algebraic Equations*, Columbia University Press, 1930.

[6] H. M. Edwards, *Galois Theory*, Springer, 1984.

[7] L. Gaal, *Classical Galois Theory with Examples, (fourth edition)*, Chelsea, 1988.

[8] C. R. Hadlock, *Field Theory and Its Classical Problems*, Math. Assn. Amer., 1978.

[9] N. Jacobson, *Structure of Rings*, Amer. Math. Soc., 1956.

[10] N. Jacobson, *Basic Algebra I*, Freeman, 1974.

[11] I. Kaplansky, *An Introduction to Differential Algebra*, Hermann, 1957.

[12] I. Kaplansky, *Fields and Rings (second edition)*, University Chicago Press, 1974.

[13] A. R. Magid, *Lectures on Differential Galois Theory*, American Mathematical Society, 1994.

[14] G. A. Miller, H. F. Blichfeldt, and L. E. Dickson, *Theory and Applications of Finite Groups*, Dover, 1961, Originally published by Wiley, 1916.

[15] E. Netto, *Theory of Substitutions*, Chelsea, 1961, Reprint of 1882 edition.

[16] J. Rotman, *A First Course in Abstract Algebra*, Prentice-Hall, 1996.

[17] J.-P. Tignol, *Galois's Theory of Algebraic Equations*, Wiley, 1988.

[18] B. L. van der Waerden, *Modern Algebra (fourth edition)*, Ungar, 1966.

[19] B. L. van der Waerden, *A History of Algebra*, Springer, 1985.

[20] H. Weyl, *Symmetry*, Princeton, 1952.

Index

Universitext   *(continued)*